Geeky-Girly Innovation

Geeky-Girly Innovation

A Japanese Subculturalist's Guide to Technology and Design

Morinosuke Kawaguchi

Translated from the Japanese
by Mark Schilling

Stone Bridge Press • *Berkeley, California*

Published by
Stone Bridge Press
P.O. Box 8208
Berkeley, CA 94707
510-524-8732 • sbp@stonebridge.com • www. stonebridge.com

Otaku de Onnanoko na Kuni no Monozukuri ©2007 Morinosuke
Kawaguchi.
All rights reserved.
First published in Japan in 2007 by KODANSHA LTD., Tokyo.
Publication rights for this English edition arranged by
KODANSHA, LTD.

Transation from the Japanese by Mark Schilling.

Book design and layout by Linda Ronan.

For image credits and references see page 254.

Printed in the United States of America.

10 9 8 7 6 5 4 3 2 1   2016 2015 2014 2013 2012

LIBRARY OF CONGRESS CATALOGING-IN-PUBLICATION DATA
Kawaguchi, Morinosuke, 1961–
  [Otaku de onna no ko na kuni no monozukuri. English]
  Geeky-girly innovation : a Japanese subculturalist's guide to
technology and design / Morinosuke Kawaguchi.
     p. cm.
   ISBN 978-1-61172-002-0 (hardback)
 1. Product design—Japan. 2. Subculture—Japan. I. Title.
TS105.K3813 2012
 658.5'752—dc23
                          2011051601

Contents

rule 4
Leave *Something* Out:
Sundome   *107*

rule 5
Connect Us with Kindness:
Communal Values   *127*

rule 6
Eliminate Embarrassments:
Modesty Pursuits   *149*

rule 7
Keep It Clean (and Healthy): Avoiding the Icky Sticky *165*

rule 8
Transform Life into Theater: Everyday Drama *180*

rule 9
Be Good to the Planet: Eco-Friendly/Eco-Product *195*

Preface: To My English-Speaking Readers

Thank you! It makes me very happy that you were curious enough to pick up my book. I am grateful that this English version is available so that the ideas it presents can be shared with more people around the world. I hope that reading it will be a fun experience for you.

The examples I give in the text are all drawn from Japanese subcultures, but my message is not limited to Japan or Asia. One universal aspect of all subcultures—no matter how much our nations, peoples, languages, art forms, and behaviors may differ at first glance—is that they are treasure chests filled with ideas that are fun to explore. The creation of so many talented artists and geeks, these subcultures all have their own originality, personality and identity. And since they are a mirror image of their own "parent" culture, they offer a rich source of hints for innovation in technology and design and in product functions and specifications.

Today the world is becoming more connected but also more diluted by broadband and globalization. The important thing for any company is to find out what differentiates their products from their competitors'. You need to know who you are if you want to innovate. So forget the benchmark of your competitors: Know yourself!

I hope this book will be a trigger and a unique guide for you on your journey to know yourself even more.

Thank you for listening, reading, and searching. I'm looking forward to hearing from you at m@morinoske.com.

Morinosuke Kawaguchi
Tokyo, Japan

Introduction to the Japanese Edition

As a technology management consultant, I help businesses assemble innovative research-and-development (R&D) teams and decide which technologies the teams should focus on.

The requests I get from companies have changed over the years. I used to field a lot of questions about how to conduct business. But lately, companies want to know what they should be making. In other words, the focus has shifted from business practices to products.

Once Japan's Bubble economy collapsed in the early 1990s, the country floundered and lost confidence for the rest of the decade. Pundits said Japan should learn from the West. Acronyms like "quality ISO," "CRM," and "ERP" became buzzwords in Japanese business circles.

Japan is still recovering from its economic slump. Most of the country is mired in doubt and anxiety, but some experts predict a revival akin to the boom from 1966 to 1970. In the past decades, Japan imported all sorts of supposed "global standard" Western management methods that focused on the "how" of business and helped improve key operational metrics, but the Japanese remained perplexed, stuck in limbo. This was borrowed advice that didn't quite fit. We kept thinking, "What should we be making now?"

To answer this question, we need to know who we truly are. What makes us tick? Where are our strengths and weaknesses? Only by answering these questions can we create a manufacturing culture that expresses our true intent.

In the field of strategic consulting, we often perform a SWOT analysis, which evaluates a situation from four perspectives: Strengths, Weaknesses, Opportunities, and Threats. SWOT analysis is a common-sense approach that focuses on one's actual ability, not abstract theory.

When I applied this analysis to Japan, I realized that two of the more positive aspects of contemporary Japanese culture were surprisingly the long-derided *otaku* (geeky) and *gal* (girly) subcultures.

The *otaku* subculture is mostly male, often obsessed with pop-culture ephemera or obscure and even perverse genres of porn, and generally considered the den of socially inept losers. The *gal* subculture features girls who dress in extreme—and often extremely revealing—"Western" styles that almost always feature heavily mascaraed eyes and dyed hair. Mainstream society sees them as delinquents or dropouts.

The center of *otaku* culture is Akihabara, a district in Tokyo known for electronics and software that has become a mecca for *manga* (comics), *anime* (animation), games, and character goods. Tourists flock to Akihabara.

Gal culture is thriving in western Tokyo's Shibuya district, the epicenter of youth fashion. *Gal* culture has pushed Japan's fashion industry from an obsession with designer brands to more grassroots fashion trends.

When cultural movements such as these start to merge into the mainstream, the Japanese tend to roll their eyes and complain about the lamentable state of society. But why can't the originality we see in *otaku* and *gal* culture contribute to the creation and manufacture of new products?

I began reflecting on this question and ended up writing this book as my answer.

It's easy to write off *otaku* and *gal* culture as mere child's play. What possible use could they be to business?

Introduction to the Japanese Edition

But that would be missing the point. These nascent youth movements need to be nurtured, just as we would nurture a child.

In fact, educating children and growing a business have a lot in common. Children come in all shapes, sizes, and abilities. Some hate studying but are clever with their hands. Others are poor at languages but love computers. Parents need to help their children capitalize on their strengths and not punish them for their weaknesses.

What if your child identifies with *otaku* or *gal* culture? I believe that parents must find the positive qualities within those subcultures and encourage their children. This may not be easy, but it is essential in any country.

Japanese society at large needs to do the same. We need to harness the strengths of the geeky and girly subcultures to help pull the larger economy out of its morass.

In this book, I argue that *otaku* and *gal* subcultures are not a race apart; we all share geeky and girly traits—even those of us who frown upon these youth movements.

Japan is a world leader. But Brazil, Russia, India,

China, and several other Asian nations are hot on our trail. To survive in this highly competitive world, we need to find a path of our own to follow.

It often takes an outsider to point out to the Japanese their inherent strengths. The Japanese tend to downplay or ignore their own talents, even though the rest of the world is well aware of them.

In the first section of this book, I examine products with distinctly Japanese characteristics. Why were these idiosyncratic products created in Japan? Why do they have such strange features? By examining the unique aspects of these products, we can better understand the culture that produced them.

In the second section, I introduce my ten rules for geeky Japanese products and give you the context to understand where these quirky ideas come from. Then I wrap up in section three by examining just what it means to be girly and geeky, and why *"otaku* manufacturing" is the key to ending Japan's economic doldrums.

The post-Bubble decades of the 1990s and 2000s are often referred to as "the lost years," a never-ending parade of failures. But that is not the whole story. During this period, Japan's creativity has been germinating and preparing to flower.

The products in this book may seem wasteful, indulgent, decadent, even idiotic at first glance. But look closer and you'll find that these so-called crazy products are pointing us in the right direction.

M.K.

part 1

Like a Girl: Childish Japanese Products

chapter 1

Seeking "Extravagant Convenience"

Even Madonna digs Japanese toilets

Sometimes even rich celebrities are blown away by the extravagant convenience of certain Japanese products. When pop star Madonna toured Japan for the first time, she told the media that her favorite product was the pre-warmed toilet seat. Imagine: Until that time, even multi-millionaire Madonna sat on a cold toilet.

Another woman impressed by Japanese toilets is my wife, a European-born American who has spent fifteen years in Japan. She was impressed with them when she first arrived in Japan and still raves about them today. She regards them as the perfect distillation of Japanese originality. What is it about Japanese toilets that impressed Madonna and my wife, Judit?

To answer that question, let's use motion analysis, a technique businesses employ to improve productivity

Japanese toilets are full of surprises.

at factories and other manufacturing sites. We'll examine chronologically the actions that occur between entering and leaving a toilet.

First the user opens the bathroom door and enters. The latest models of Japanese toilets are equipped with infrared sensors that detect movement and automatically raise the toilet lid, like an oyster opening its shell.

This function may seem unnecessary at first glance. But since the toilet lid doesn't need to be touched, it stays clean and so do our hands. Also, elderly or disabled users don't have to bend down and small children don't have to extend themselves over the toilet bowl to flip up the lid. Finally, since the lid also closes automatically, it conserves energy by capturing the heat of the pre-warmed seat.

A typical control panel on a Japanese toilet. Which button to push?

©Armin Kübelbeck

Japan's toilets used to be of the squat variety. It wasn't until the late 1980s that the production of Western-style toilets surpassed that of Japanese squat-style toilets. And yet it was the Japanese manufacturers who came up with the pre-warmed seat. I feel sorry for the millions—no, billions—of people who still shudder as they lower themselves onto their chilly toilet seats.

Naysayers will argue that these extravagant toilets waste electricity at a time when manufacturers should be going green. But manufacturers are addressing those concerns; the latest models heat rapidly to reduce electricity usage.

After sitting down, the user eliminates. At this stage, a Japanese toilet will emit a flushing sound to mask the sounds of elimination. Toto Corp. introduced this feature

Seeking "Extravagant Convenience"

in 1988 with its Otohime (Sound Princess) device, which can still be found in many office toilets. Now women no longer need to flush twice to mask the splashing sound when they pee. This saves water while also reducing embarrassment in the toilet.

I once asked my wife if this embarrassment was universal or strictly a Japanese thing. "Westerners feel embarrassed too," she told me. "Even in my elementary school, the girls were embarrassed to use the toilet at school. Now that I'm used to the flush simulators in Japan, I get tense when I have to use a public toilet abroad."

That means the flush simulator's potential market is global in scope. We should be looking for more potential hit products like Otohime. Japan has talented designers who are tuned into such widely shared (if unspoken) feelings, forward-looking companies that are eager to bring such products to market, and consumers with a highly developed sense of shame who are eager to embrace those products. How can we lose?

Another reason people flush the toilet while eliminating is that they are embarrassed about the smell drifting out of their stall. The latest toilets from Japan come with powerful deodorizers, which eliminate smells catalytically, making replacement parts and deodorants

Press the Sound Princess while relieving yourself, and the sound of flushing water will drown out other embarrassing sounds.

unnecessary. More and more toilets also have ventilators with automatic timers. After the user exits, the ventilator fan keeps running for a few minutes, before shutting itself off. Such is the Japanese penchant for detail.

Automatically folding toilet paper

Built-in bidets are standard issue in so-called Japanese shower toilets these days. The cleansing spray feels nice while also reducing the amount of toilet paper we need to use.

Toto launched the first bidet-style spray cleaner, called the Washlet, back in 1980. Three decades later, the rest of the world is starting to come around: About 10% of Toto's sales come from overseas now.

A trip to the bathroom in Japan is full of little luxuries like folded toilet paper.

Hotels in Japan fold the toilet paper ends into neat triangles. That little extra touch reminds me of a man in formal wear with a triangle of handkerchief peeking out of his chest pocket. The idea of folding toilet paper in triangles came from hotel chambermaids, who used it to signal guests that the toilet had been cleaned. It became a common practice in Japanese luxury hotels in the 1960s with the advent of roll-type toilet paper and is now being adopted by luxury hotels around the world.

Kei Automobile Parts Manufacturing had the busy chambermaid in mind when it developed a toilet-paper holder called Mervoi that folds the paper with a single-lever action, reducing work for hotel staff.

Once the toilet user has finished his or her business, it's time to flush. With a Japanese toilet, you have a choice: a big flush or a small one. In fact, this has been a standard part of Japan's toilets for a long time. Pull the lever to one side, and you get a small amount of water, enough to flush liquid waste. Pull it the other way for solid waste. The *dai/sho* (big/small) lever is simple, easy to use, and environmentally friendly, and yet it still hasn't caught on overseas.

The *dai/sho* lever is a good example of what Nobel laureate Wangari Maathai called Japan's culture of *mottainai* (a hard-to-translate word that refers to lamentably wasteful situations, actions, or things). The latest toilets combine this function with cutting-edge electronics to create an auto-flushing feature that measures the water required for flushing and flushes as soon as the user stands up.

Now it's time for the user to wash her hands. The newest toilets dispense water and liquid soap when the user waves her hand in front of a sensor. Then the electronic jet-air dryers dry her hands, which means she doesn't have to touch the toilet with her hands at all!

Sure, the Japanese passion for cleanliness borders on obsession, but it has also led to some very creative devices that allow us to use public restrooms in a sterile fashion. It reminds me of the way surgeons enter an operating room by holding up their hands and pushing the door open with their backs.

From geeky minds come luxurious functions

Here's another wonderful device for the lavatory: sanitary napkin packages that open silently (made by such firms as Unicharm and P&G Japan). Let me explain this to men who have no idea what I'm talking about. Sanitary napkin packages used to make loud tearing noises when women opened them, causing embarrassment for some. Unicharm and P&G used more flexible packaging material to develop the silent packs. Like the Otohime flush simulator, this product was made for people who feel shame in such situations. I admire the people who not only unearthed these hidden yet real needs, then designed products to meet them.

The advanced Japanese toilet is not all high-tech and gadgets. A lot of the ingenuity comes from a high sensitivity to consumer needs. You might call them "enlightened features."

Japanese often say that if you are planning to work for a company, check out its toilet first. The same is said about prospective marriage partners. That's because the

The latest high-tech toilets, like this wireless one, come with panels that seem like they belong in a cockpit.

toilet most graphically expresses the true state of a company or family.

Let me sum up the points I've made so far:

1. Japanese place a high priority on convenience. Manufacturers try to automate every aspect of the toilet experience.

2. Japanese have a passion for cleanliness. Users of Japanese toilets often do not need to touch a thing, and when they do have to press a button, it's typically covered in an antibacterial coating.

3. Japanese are relentless in their pursuit of comfort. One advantage of shower toilets is paper conservation, but for Japanese users a more

Seeking "Extravagant Convenience"

important benefit is the pleasant shower-like sensation they provide.

4. Japanese are sensitive to shame. This has resulted in products such as the Otohime flush simulator and silent sanitary napkin packages.

5. Japanese are considerate of others. This is reflected in catalytic deodorizers and the chambermaid's practice of folding the edges of toilet paper rolls.

6. Japanese are concerned about the environment and waste. The big/small flush levers help conserve water, as do new high-tech mechanisms on water-saving toilets that flush just half the amount of water—about 3.8 liters—of older models.

While many of these creative additions to the toilet may not be part of the toilet's basic function, they are more than frivolous luxuries. Japan's highly developed—or some would argue "overly developed"—sense of luxury and a geeky attention to detail sharpen the Japanese product and make it unique in the global marketplace.

Wash your hands in the water destined for the next flush to reduce water waste.

The huge potential of Japan's "girly" culture

All products and services have a life cycle or an expiration date. They go through an embryonic period in which they are created, then an expansionary period in which they acquire a range of functions. Then there's a period of maturation in which competition and supplementary features come to the fore.

Over time, services and features that were all the rage because of their novelty and originality become commonplace. At this stage in a product's life cycle, a Western company may wonder why it should continue making the product. Why continue churning out the products just for the slim profit margin? Why not sell off the business and cash in?

For many Japanese companies, the real struggle

Nearly a century ago, Marcel Duchamp argued that the toilet was a work of art. Today, Japanese manufacturers have proven his point.

begins at the point where the Westerner is ready to sell. Instead of cashing in, a Japanese company will try to improve the product, adding new features and trying to breathe new life into the commonplace item.

The basic function of a toilet is to eliminate, flush, and finish. Japanese makers have added a multitude of features, turning the toilet into a fully outfitted ship in your bathroom! You could even argue that Japanese toilets are a work of art (it worked for Marcel Duchamp!). Products that go through this creative process have the power to hook people, regardless of national boundaries.

But when a product has reached maturity, developers do need to ask themselves whether it is overripe and starting to rot—whether all those extra features add up to overindulgence.

Like a Girl: Childish Japanese Products

We'll come back to that point. But first, let's look at the Japanese love of luxury from an individual perspective.

Some people become preoccupied with their appearance and fixated on certain brands they think will help them improve their look. Some prefer *kawaii* (cute) products, others fashionable designs. Health benefits and safety concerns can also influence their choices. Today a preference for only basic functions has become a kind of luxury. It's like saying, "A phone is a phone. I don't need all these things like text messaging and built-in cameras."

A sense of luxury brought pre-warmed toilet seats to Japan (after all, you can always do your business on a cold seat). This sensibility has long been embedded in Japanese culture. It's part of who we are.

Japanese can behave childishly or be painfully shy because they care too much about how others perceive them. They cultivate "inscrutability" by leaving a gap between their public face (*tatemae*) and private feelings (*honne*). Many Japanese perceive this as a liability in an increasingly global society. They'll argue that Japanese need to express themselves freely and directly instead of always being so concerned with what others think.

Really?

It's been over 150 years since Japan abandoned

Once you sit on a warm toilet seat, cold seats will seem positively barbaric.

its policy of enforced national isolation. And ever since that policy changed, Japanese people have been beating themselves up over their childish ways and cultural insularity. But I ask you: If we had become more rational and international—more adult in the Western sense—would we have made such luxurious toilets?

I doubt it.

If we see the rational, internationalized man as an exemplar of Western adult male values, then overindulgent, luxurious Japanese products express the opposite—the "girly" culture, if you will.

Instead of imitating the West, we should play to our strengths. While it's admirable to want to compensate for our weaknesses, we should also confidently exploit our girlishness.

Like a Girl: Childish Japanese Products

Don't get me wrong: Japanese are very good at macho stuff too. We can use ultra-high-speed computers to send intercontinental ballistic missiles to their targets, and the world's fastest supercomputer happens to come from Japan. But manufacturers will be better off if they head down the path of innovative Japanese toilets, which symbolize the power of manufacturing based on true Japanese sensibilities.

"Girly" products can spur Japan's growth in this century every bit as much as, if not more than, the "manly" technologies.

chapter 2

Making Friends with Tools and Machines

The Hello Kitty sticker that surprised my wife

An essential quality for strengthening Japanese manu-facturing is its childishness. When my wife first came to Japan, she often made fun of all the manga and anime characters she saw in public, calling them "childish." Since then, however, she has been completely won over by Japanese "character culture," and our house is full of *kawaii* (cute) paraphernalia. She was recently surprised when she saw stickers printed with the likeness of Hello Kitty, a popular cute character, on commuter trains in Tokyo. If it had been the ordinary Hello Kitty, who never says a word, she would not have thought it unusual, but the Hello Kitty on the sticker was saying, "Don't get your fingers stuck! Stay away from the door!" with a bandage wrapped around one of her fingers.

This sighting of Hello Kitty in the urban landscape

surprised my wife, although she has lived in Japan a long time and understands its culture. She was shocked at this extreme example of Japanese anthropomorphism—the tendency to give animals and even inanimate objects human qualities. No Japanese person would have been taken aback by this sticker, though. Japanese accept these kinds of characters without reservation in nearly any situation. It's this anthropomorphic sensibility that allows them to have such close affinity with inanimate manufactured goods.

Putting kitchen knives and sewing needles to rest

Perhaps you've heard of the Japanese custom of holding funerals for kitchen knives and bent sewing needles. Japanese develop strong emotional attachments to inanimate objects. They come to see them as inhabited by spirits, even if they are made of metal.

The Japanese word "*dogu*" and the English word "tool" have quite different nuances. In the West, children are allowed to develop emotional attachments to objects, but adults are expected to grow out of this fantasy world and detach from the inanimate. Few Westerners get a warm, fuzzy feeling when they hear the word "tool." In Japan, however, adults don't have to detach their emotions from their tools, thus the term "*dogu*" continues to carry warm, affectionate connotations.

When a tool—be it hardware or software—is created, a barrier exists between the human and the tool. Stress occurs when a person operates a machine or handles a tool for the first time because he or she has to learn how to use the tool. The tool doesn't explain itself, after all.

Machines that scoot closer to humans

Increasing an engine's horsepower or optimizing the operational speed of microchips is an important task. But when designing tools, a higher priority is to make the interface naturally accessible to humans. Otherwise, the person won't relate to the tool, and it will end up getting tossed aside.

It's not enough to study mathematics or physics; designers also need to understand the human using the tool.

Japanese culture has a deep-rooted tendency to anthropomorphize the inanimate world. The aforementioned funerals for kitchen cutlery and sewing needles are a perfect example of seeing spirits in the inanimate. And these sentiments form the foundation for today's "character culture," including manga and anime. So much of Japanese culture is anthropomorphized that the Japanese today seem to be living in a theme park or fairyland.

Take, for example, the faceless cellphone of yore. The cellphone's original function was to serve as a long-distance communication tool. This was later supplemented with texting and image-exchange functions, making the cellphone a multimedia device. Then came music, personal finance, and satellite navigation functions, further transforming the phone into a multi-feature tool. The cellphone of the future will evolve into a richly personalized concierge, secretary, and butler for all spheres of daily life.

When I explain this to Japanese youths, they put it much more succinctly: "You mean we're going to have a Medama Oyaji."

Medama Oyaji is a character in the famous *yokai* (mythical beings and monsters) comic *GeGeGe no Kitaro*. He's the father of the protagonist Kitaro but he looks like an ambulant eyeball. Small and portable, he usually hides in Kitaro's hair. When his son is in trouble, he whispers advice in Kitaro's ear—just like a navigation aid.

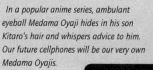

In a popular anime series, ambulant eyeball Medama Oyaji hides in his son Kitaro's hair and whispers advice to him. Our future cellphones will be our very own Medama Oyajis.

©Mizuki Pro.

When his son is feeling down, Medama Oyaji dishes out encouragement while seated in a tea bowl—a one-man counseling and coaching service. He knows all of Kitaro's strengths and weaknesses, and can provide optimal solutions in real time. In the 21st century, we are much more likely to see science create a Medama Oyaji than an Astro Boy, the nuclear-powered robot hero of Osamu Tezuka's famed 1960s cartoon.

From the room-sized mainframes of the 1970s to the personal computers of the 1980s and the mobile communications devices of the 1990s, computers have followed a path of miniaturization, with each development bringing them closer to human scale. Now pocket-sized products are gaining popularity.

Some scientists predict that after miniaturization

Like a Girl: Childish Japanese Products

has made products wearable, like clothing or accessories, the final step will be a direct link to the nervous system. In clinical trials, researchers have already connected electrodes to the brains and nervous systems of disabled patients, allowing them to control artificial limbs and robotic arms through the power of thought.

Machines are closer to us than ever. Can Japanese makers close the gap even more? Perhaps they can bring machines *psychologically* closer to humans, tailoring them to the user's individual personality. Then we'll all have our own Medama Oyaji.

Why do we like mosaic patterns and aligned numbers?

What attracts people to products? What do people want products to do for them? Answering such questions is the first step in solving the problem of man/machine interfaces.

A product's branding and functions are certainly important. But also important, and frequently overlooked, is artistic quality. People are drawn to and derive pleasure from excellent works of art.

When Japanese people talk about marrying art

and technology, they usually begin and end the discussion with *kogeihin*, or traditional Japanese craftwork. Craftsmen who have spent years perfecting their skills can elevate their output to the realm of art. Today, small industrial workshops in Japan have developed sophisticated techniques rivaling those of traditional craftsmen. The artistic appeal in the work of the expert craftsmen and the highly evolved industrial workshop lies in the melding of function and beauty.

People known as "media artists" bring artistry to the making of intangible software. When we talk about art, we usually think of *Venus de Milo*, *Mona Lisa*, Beethoven's symphonies, Oribe's tea bowls and Dostoyevsky's novels. There's no room for the media artist in this traditional definition of art.

The media artist ponders why we have the urge to look at our digital watches at the 11th second of the 11th minute of the 11th hour. Or when a child walks on a mosaic tile floor, they ask why he or she always wants to step on the center of the pattern? They notice that traffic lights tend to loom haughtily above us and wonder if they could be more humble as they turn red and tell us to stop in our tracks. Media artists know that people are drawn to objects by the same mysterious mental process that makes them look at their watches when the numbers match up.

Media artists in Japan are trying to minimize barriers to man/machine interfaces. In future chapters, we'll take a closer look at their efforts and at the Japanese penchant for attaching themselves emotionally to their tools and products.

part 2

Ten Geeky-Girly Rules of Japanese Products

Make the Object (Almost) Human: Anthropomorphism

Star-filled eyes as headlights

Of all the machines we normally purchase, a car ranks as the most complex. This machine converts gasoline-fueled combustion into thousands of rotations per minute, propelling a weight exceeding 1 ton at speeds of 100km/h. Within the comfortable interior of the car, it is hard to imagine the drama going on under the hood.

But the success of the Japanese car industry is not due solely to superior gas mileage, reliability, and other performance factors. Continual improvement in mass-production technology has also made Japanese cars highly cost effective. This combination of superior performance and cost effectiveness, developed over many years, has given Japanese cars a strong reputation.

Most carmakers produce reliable cars when it comes

Make the Object (Almost) Human: Anthropomorphism

to basic performance. How do we tell one from the other? The key is for carmakers to design vehicles that fit their corporate strategy.

Products can be sold on the basis of the images and feelings they evoke in users. A good example is cosmetics, where the actual performance of the product is often not the prime reason to buy it. And while cars are the result of sophisticated engineering processes, they can also be an emotional product.

In car design, the most important area for expressing originality is the front, or "face" of the car. Headlights are the car's eyes. Porsche once led the way in headlight design with its influential sharp lines. The type of headlight most in vogue now, however, is designed to twinkle. An array of lights like beam guns is encased in a transparent cover to which is attached a polyangular reflective panel called a multi-reflector. Instead of the streaky cuts on the cover lens of conventional lights, designed to maximize light focus, there is a clear view into the interior. HID (High Intensity Discharge) bulbs produce a super-bright blue-white beam. All this serves to radically transform the look of the car's face.

These new headlights were first installed in the Nissan Cima in 2001, symbolizing Nissan's then upward

The latest headlights from Japan mimic the wide-eyed look of manga and anime characters.

©Tezuka Production.

momentum and renewed vigor, while considerably sharpening the car's expression. A hit with buyers, they are now rapidly being adopted by carmakers worldwide.

Most Japanese have seen these giant twinkly eyes before. Large twinkly eyes have long been the standard for characters in manga for girls. The originator of this twinkly look is said to be the character Princess Sapphire in Osamu Tezuka's 1950s manga for girls, *Princess Knight*. The abnormally huge eyes have created a huge visual contrast between Japanese manga and conventional Western comics.

The princesses in Disney cartoons, for example, have eyes like those of real-life beauties. Similarly, Western car companies such as Ferrari and Citroen are engaged in a quest for elegant humanlike headlights. As a result, their

Make the Object (Almost) Human: Anthropomorphism

beautiful monocular headlights are markedly different from Japanese headlights.

One hot innovation is a new type of steering headlight system. This intelligent system, in which the steering wheel and headlights are connected, goes into action when a vehicle is turning a corner. Steer to the right, and the lights turn in the same direction, lighting the way for the driver.

Yet another popular Japanese headlight resembles the Zaku mobile suit from the popular *Mobile Suit Gundam* TV animation series (broadcast from 1979 to 1980 in Japan). Zaku was more popular than the protagonist and had eyes that realistically shifted from left to right.

In a country where the car industry is so vital to the economy, it's intriguing that the "eyes" of a car's face are like those of manga characters. Perhaps the era the car designers grew up in left an unconscious imprint on them.

Japanese comics have evolved greatly since the war and are now read around the world. Members of the first generation to have grown up with manga are becoming executives of big corporations. Manga's influence on Japan's manufacturing industry will no doubt continue to grow, generating new products and concepts.

Motorcycle faces, Gundam, and *kumadori*

Do you think I'm overstating the link between manga and manufacturing? Let me try to sway you by looking at the design of the front end of motorcycles in Japan.

Many motorcycles and scooters are equipped with twin headlights. The front of the bike is a streamlined scowl with two "eyes." Larger motorcycle and scooter models are mostly of this type.

Twin lights that resemble scowling, intimidating eyes are becoming standard for psychological reasons that go beyond giving machines a sporty, masculine look.

People tend to have strong reactions to expressions of anger. If a teacher walks into a classroom with a scowl on his face and glares at the students, they will quiet down immediately. Likewise, if a motorcycle or scooter has an angry face, the driver in front will immediately notice this in his rearview mirror, as will pedestrians. This is important for accident prevention; if you're a bit scared, you're likely to hesitate before doing something dangerous.

Honda employs neurologists to conduct research in this area. They use fMRI (functional magnetic resonance imaging) to investigate brain activity in reaction to human faces. Subjects were shown samples of motorcycle

The headlights of motorcycles resemble the scowling eyes of a Kabuki actor.

fronts, and neuroscientists measured their reactions. The design that left the biggest impression on subjects was the front designed as an angry face. Honda adopted the scowling front in an attempt to reduce accidents.

Kabuki actors use a makeup technique called *kuma-dori* to draw thickly marked eyelines—they make a strong impression on audiences. Those same *kumadori* eyes are used on crime prevention posters in Tokyo with the tagline: "Somebody's watching!" The aim is to give potential criminals the willies, as if they are under perpetual surveillance.

The *kumadori* effect is also found in robot anime. Consider Gundam's face, for example. The effect in robot anime is similar to the effect of the scowling headlights on a motorcycle.

A recent prototype by the Honda research team features a round, crystalline lens in the middle of the light beam, which resembles a human pupil. By suggesting a pupil as well as the outline of an eye, the overall effect becomes stronger.

Cosmetic contact lenses with a clearly defined iris outline are trendy now. The outlined iris gives wearers attractive, wide-open eyes. But emphasizing the eyes in such a matter can make intimidating faces even scarier. BMW and the Nissan Fuga already use this look in their headlights.

Japanese also humanize machines in friendlier ways. Humanoid robots like Honda's Asimo have been readily accepted in Japan as almost human. This attitude of acceptance has been shaped by manga and anime, which depict robots such as Osamu Tezuka's iconic Astro Boy in a positive light. Even before the advent of manga and anime, the Japanese tended to regard non-human things as human and thus easily accepted the personification of objects.

Funeral rites for needles, chickens, and germs

Japan is a polytheistic country. Some say it has as many

as eight million gods, many of which reside in inanimate objects. Our ancestors naturally formed emotional attachments to these objects, and we still do today.

For example, when disposing of decorative *hina* dolls (commemorative dolls for girls) and ornamental armor (for boys), some people hold funeral rites. Perhaps non-Japanese can understand this sentiment. But the Japanese go even further. As I mentioned earlier, they'll hold last rites for kitchen knives and scissors. I think there are few people outside of Japan that can form an emotional attachment to a pair of scissors!

The ultimate expression of this mindset is the last rites for needles. The ceremony involves sticking a blunt or broken needle in tofu or *konnyaku* (devil's tongue). These soft foods are used out of compassion for the needle, which led a torturous existence of being stabbed into other hard objects. Doesn't that send-off for the needles choke you up just a little?

Of all the tools we use, needles have to be among the hardest to become emotionally attached to. But even that sliver of metal with a hole in one end has a spirit.

There are no last rites for toothpicks in Japan, but Japanese toothpicks have two grooves carved in one end. Why the two grooves? One explanation I like is that you

When a needle is no longer useful, it is buried in a soft piece of tofu or konnyaku *as thanks for all the hard work it did in its lifetime.*

can break off the toothpick at the groove and use it as a pillow for the toothpick. Grooves are carved into toothpicks so that Mr. Toothpick can have a pillow to sleep on.

Kentucky Fried Chicken has franchises in 109 countries, but only in Japan does it have a Broiler Appreciation Ceremony to commemorate dead chickens. Japanese KFC executives go to a Shinto shrine to bow their heads to chicken-*san*. The Japanese also commemorate the deaths of livestock and lab animals such as mice and rabbits. Until recently, the practice of commemorating dead insects was common. And the Japanese people are probably the only people in the world to have built a tomb for dead bacteria to show appreciation for the microbes found in food, medicine, detergents, and other everyday items.

Make the Object (Almost) Human: Anthropomorphism

Even Japan's Self-Defense Forces have cute mascots. Meet Prince Pickles and Princess Parsley.

©Taro Tomonaga.

The world of *-tan*: anthropomorphizing everything

In Japan, anthropomorphic mascots are everywhere you turn. They are such a natural part of the landscape, that they're often overlooked. A good example of this is Hello Kitty warning train commuters not to get their fingers stuck in the automatic doors. These mascots give us insight into Japan's anthropomorphic culture.

It's no surprise that mascots are used to target children in ad campaigns for amusement parks, stationery goods, and toys. In Japan, though, mascots are attached to far more straight-laced, adult enterprises as well. Take, for example, Prince Pickles and Paseri-chan (Miss Parsley), image characters for Japan's military, the Self-Defense Forces. Their names suggest that the SDF is a group of

The police departments of all forty-seven prefectures in Japan have mascots. Here are, counterclockwise, Joshu-kun and Miyama-chan of Gunma, Fukuboshi-kun of Fukushima, and Fu-kun and Kei-chan of Osaka.

helpful, modest, self-deprecating folks. The Tokyo Police Department also has its own mascot: Peepo-kun, named after the wailing sound of police sirens (*peepo-peepo* in Japanese). Every one of Japan's forty-seven prefectures has its own variation of Peepo-kun representing its police force.

Conservative organizations that embody state power use cute mascots to soften their authoritarian air. In a similar vein, public ad campaigns for preventing AIDS, stopping gangster violence, and tackling drug addiction employ cute characters who obscure the harsh realities of their subject matter. The cover of the manga version of the 2005 White Paper on Defense shows a girl holding down a lightly billowing skirt to hide her panties, creating a *chirarism* effect. ("*Chirarism*" is a Japanese-English term

Make the Object (Almost) Human: Anthropomorphism

that means to fetishize briefly glimpsed objects of sexual desire.)

An even more intense form of mascot obsession can be found within the *moe* culture of the *otaku*, who have created the world of "*-tan*." ("*Moe,*" in the *otaku* world, means friendly or affectionate feelings toward an object of fetishized desire, be it a café waitress in a frilly maid costume or a figurine of a female anime character.)

Japanese sometimes call people "*-chan*" (an affectionate suffix usually reserved for young girls and cute animals), but *otaku* who are into *moe* have started to use "*-tan*," which is a baby-talk version of "*-chan*." This is another big step in the anthropomorphization process.

That process was once directed mainly at animals, insects, and other living things. Mickey Mouse and Bambi come to mind. Machines such as cars and steam engines are also easy to imagine as human characters. Just think of Thomas the Tank Engine, for example. But in *moe* culture, "*-tan*" transforms all phenomena and abstract concepts into *bishojo* (pretty young girls). Even traditionally macho objects such as weaponry (Battleship Yamato-*tan*, Messerschmitt-*tan*) and adult products like cigarettes (Marlboro-*tan*, Mild Seven-*tan*) get the *-tan* treatment.

Even concepts transform into pretty young girls.

Even software can be turned into pretty young girls through the -tan process.

There is Windows-*tan* and Winny-*tan*, for example, representing computer software, and Article Nine-*tan*, which turns the peace clause in the Constitution into a cute girl. Or perhaps you'd be more attracted to Anti-Monopoly Act-*tan*? Physics geeks may take a shine to Kirchoff's Law-*tan*, and Middle East experts will doubtless gravitate toward Afghanis-*tan*. Almost all of these characters are mildly eroticized, a trademark of *moe*.

It's impressive how skillfully the makers of these characters codify the appearances and meanings of their subjects. This is the new approach: *moe*-anthropomorphization. And it's largely done by very skilled amateurs, not professional artists or designers. Some objects have many different -*tan* versions because many people are inspired to create them.

Make the Object (Almost) Human: Anthropomorphism

Turning Article Nine and spaceships into pretty young girls

Of all the train-based *moe* characters, the most popular is Fastech-*tan*. Japan Railways East created the prototype Fastech 360 in 2005 as a next-generation bullet train capable of speeds up to 360kph (223mph). When Fastech slows down, triangular flaps unfold from the top of the train, acting as brakes. Fastech caused a stir among *otaku* who thought that the brake flaps resembled cat ears, so they created Fastech-*tan*.

Fastech-*tan* is a teenage girl color-coordinated to resemble the real Fastech. She has green hair and wears a white and blue miniskirt. Her white knee-high boots end in catlike feet; white gloves cover her paws.

By contrast, Thomas the Tank Engine is still shaped like a tank engine, his only human feature being his face. Fastech-*tan* resembles a girl from any vantage point; only her costume reflects what she's based on.

Constitution Article Nine–*tan* is a girl dressed in a sailor-style school uniform. She sits in a feminine posture, with her knees bent inward, and whispers, "Please, no fighting . . . " Article Nine contains a clause that forbids Japan from engaging in war, so her plea makes sense.

Winny was a file-sharing software that became a

huge social phenomenon a few years back. The popularity of the software and its dubious legal standing led to the arrest of the developer. Winny-*tan* is a lost-looking little girl shouldering a huge knapsack full of precious files. She roams the Internet aimlessly.

Through their mannerisms and fashion, the girls all reflect the background and meaning of the concepts they represent.

Saito Kan, author of *Gijinka Tan Hakusho* (The White Book on Anthropomorphism and *Tan*, published by Aspect), explains the world of –*tan* this way:

> Normal anthropomorphization and *moe* anthropomorphization are two very different things. There are even *moe* personifications of (popular anime characters) Pikachu and

Make the Object (Almost) Human: Anthropomorphism

The spaceship Hayabusa becomes a blue-haired girl.

Anpanman. This is an anthropomorphization of the anthropomorphized. To put it another way, it's double anthropomorphization or "meta- anthropomorphization." . . . For instance. It's hard to feel *moe* toward the actual Nausicaä of the Valley of the Wind [the heroine of a famous post-apocalyptic manga and anime by Hayao Miyazaki], but it might be possible if I lower her age and turn her into Nausicaä-*tan*.

Of all the –*tan* characters, the one that most captured my attention was a little girl called Hayabusa-*tan*. Hayabusa is a spaceship that landed on the asteroid Itokawa, 300 million kilometers from Earth, in 2005. It used a unique thrust mechanism called an ion engine and was the first in the world to make use of the so-called earth swing, in which a spacecraft harnesses the Earth's gravitational pull to accelerate. This spaceship is a badge of pride for Japan and a crystallization of the technological innovation of the Japan Aerospace Exploration Agency (JAXA).

In June 2006, *Science*, the most respected science magazine in the U.S., ran a special seven-article feature on Hayabusa. This was the first time research from Japan was featured in the magazine. The photographs taken by Hayabusa have contributed greatly to our understanding of the history of our solar system.

Less than half a year after the news of Hayabusa's successful landing, Hayabusa-*tan* appeared on the Internet. She has a sexy look, but antennae and solar panels protrude from her body.

Hayabusa-*tan* carries JAXA's ion engine on her back like a school bag and communicates with Earth via a handheld laptop. She's said to be going on her first shopping trip all by herself. She is portrayed as injured but heroic, reflecting the real-life Hayabusa, which battled fuel leakage and repeated control issues to return safely to Earth.

The spacecraft and the fictional girl cut contrasting

figures. The spacecraft is symbolic of Japanese culture's sterner side. Its innovative technology could be transferred to weaponry. Its ability to land on a rock just 500 meters wide at a distance of 300 million kilometers puts any intercontinental ballistic missile to shame.

Hayabusa-*tan*, on the other hand, is not aggressive at all. This ability to produce a character related to an object and yet in some respects the polar opposite of that object is a skill to take pride in. It reflects a very Japanese sense of balance and refinement.

In 2006, the city of Tsukuba, where JAXA is headquartered, was connected to Akihabara (the capital of -*tan*) by the Tsukuba Express train line. This was a deeply symbolic event. Now that researchers in their lab coats in Tsukuba are connected to Akihabara *otaku* glued to their computers, we can expect fresh sparks of inspiration.

Multilingual vending machines

When you think of life with robots, what do you imagine? I bet most people picture living side by side with bipedal robots like Honda's Asimo. In reality, life with robots is already here. Of course, today's robots are often single-function types.

Need a Buddhist amulet in a hurry? Japan's got a vending machine for that.

Visitors to Japan are often surprised by the huge number of vending machines. There are said to be six million vending machines in Japan. That's one machine for every twenty-five people, making Japan the hands-down vending-machine capital of the world.

One of the reasons there are so many vending machines is that they sell just about anything: cigarettes, alcohol, flowers, sushi, underwear, and pornography are just some of the staples. Some machines even accept ¥10,000 (about $125!) bills. For a host of reasons, it would be unimaginable in other countries to leave these machines out at night.

These vending machines are a type of robot; some of them are equipped with voice functions, a phenomenon seldom seen outside of Japan.

Make the Object (Almost) Human: Anthropomorphism

Canned-coffee maker Dydo Drinco's newest vending machines are positively chatty. They're also multilingual, speaking Japanese, English, Chinese, and Portuguese, and even delivering messages in regional Japanese dialects. These machines can do very respectable Kansai (western Japan), Nagoya (central), and Tsugaru (northern) dialects. The machines can recognize time, places, and even occasions, and will adjust their speech patterns accordingly.

For instance, a Dydo Drinco machine can say, "Work hard after lunch!" and "Sorry, I'm out of change" in a slangy Osaka accent. It can also chitchat. It can monitor temperatures, so you may hear one saying on a particularly sultry day: "The weather's starting to turn warmer, isn't it?"

A first-time user may feel a little bashful about being chatted up by a vending machine. You become a "black belt" user once you are able to casually nod at the machine's remarks and accept its can of coffee with practiced ease.

Advances in computer and voice-recognition technology will enrich a vending machine's vocabulary while also helping it to understand human speech. In the future, your local vending machine may have a meaningful conversation with you. Also, as sensor functions become more sophisticated, vending machines are likely to be able to recognize repeat customers or see that someone is sweating profusely, for example, and make an appropriate comment.

Vending machines may even be able to dispense advice: "There's no need to rush. Your train won't leave for another two minutes." The user may say, "Oh really? So I have some time. Thank you!" To which the vending machine may reply, "You're welcome. My name is Yuko. Please come again." Suddenly, the user's heart is beating just a little faster.

The growing world of humanoid robots

Vending machines and household electrical appliances don't really seem like robots because they don't move and have no arms and legs. But Japan is full of moving robots too.

When I say "moving robots," I mean industrial robots. As many as forty-five percent of the world's industrial robots are in Japan. By comparison, the Japanese own only about eight or nine percent of the world's cars.

Humanoid robots compete for the RoboCup, an all-robot soccer tournament.

Most industrial robots are factory robots repeating the same motions over and over in a production line. But in the near future, robots will be working in amusement parks, hospitals, care-giving, disaster relief, maintenance work, and high-risk jobs as well as handling everyday chores.

Young Japanese are making this future a reality by participating in the many robocon, or robot contests, where individuals, school clubs, and hobbyist groups come together to build their own robots. Many of the participants in these competitions want to become robot designers in the future.

These amateur robots race each other on rough roads, balance balls and boxes, and test their skills in soccer games and boxing bouts. But they also compete in tests of mental (that is, programming) prowess like *shogi* (Japanese chess) and chess.

In Japan, the robocon held by public broadcaster NHK has led to two hundred smaller robocons across the country. Outside Japan, the most prestigious robocon is the RoboCup, held by the Swiss nonprofit International Robo Cup Committee. This competition was started in 1997 at the behest of Japanese engineers.

The most challenging part of the RoboCup is the Humanoid League, which consists of bipedal robot soccer teams. Beginning with the 2004 games, when the technology first made proper games possible, Team Osaka, backed by electronics developer and retailer Viston Corp., won three times in a row.

The smaller robocons throughout Japan offer ample opportunity to see precision-made robots in action. Some of these humanoid-type robots are finding their way into the marketplace, and sales have been increasing exponentially.

Japanese reality shows occasionally feature battling robots. Watching a father in his forties (who probably belongs to the first generation of anime viewers) stay up all night creating a robot and then look on as his daughter uses the controls to make the robot deliver

You don't have to pick up after robotic pets like Sony's Aibo.

a crushing kick to its rival is like revisiting a classic anime.

The original standing robots were Honda's Asimo and Sony's Aibo. What do these two companies have in common? Both have a technology-based blueprint for corporate success as well as a romantic story as startups rising from the ashes of early postwar Japan.

These companies did not invest heavily in robot design projects with the aim of directly applying the technology to industrial uses. Instead, they created robots because the job looked fun and stirred their desire to create.

Aibo and Asimo are about pleasure first, profit second. Western businesses that focus on short-term gains will never produce such robots.

American robots built for desert combat

American versions of these robocons seem stripped of the fun and pleasure we find at Japan's competitions. The American participants are more profit driven; they're not in it just to have fun as the Japanese are.

In October 2005, the Defense Advanced Research Projects Agency (DARPA) held its second off-road rally for robot cars. The Grand Challenge, as it was called, was a competition in which unmanned robot vehicles raced from Los Angeles to Las Vegas across 210km of desert in intense heat. The contestants were not shown the course ahead of time. As they sped toward the goal, the vehicles had to plot their own courses while monitoring road conditions.

The race takes place on bumpy roads. The robots are not bipedal; they resemble pickup trucks. The Stanford University team won the race and the $2 million prize with a time of 6:53:58. The competition was intense, a dog-eat-dog atmosphere. If an Asimo had wandered onto the course, it would have been smashed to bits in a matter of seconds.

This robocon was designed to simulate Middle Eastern warfare, where armored vehicles traverse a desert landscape. The robots were designed for kill-or-be-killed situations. American robots tend to be of this nature, providing a shield for U.S. soldiers. Japanese robots, on the other hand, tend to be born from curiosity and a sense of fun. The difference in Japanese and American robots points to the different national policies and mentalities of the two countries.

The Japanese approach is hardly frivolous. Rather, it gives the Japanese an advantage when it comes to solving the problem of perfecting the man/machine interface. It's why we design headlights to look like human eyes and can converse with vending machines. With this sensibility, Japanese engineers are more likely to lead the way in robot-human cohabitation.

Pursue Personalization: Making It Your Own

Forming attachments to our tools

The Japanese are quick to personify objects, develop an emotional attachment, and give the object a special status.

There are two types of what I call "personally customized products," or products tailored to individual tastes and uses.

The first are tailor-made to individual specifications from the ordering or purchasing stage. The second are products whose rough edges have been worn down through long years of use and have come to perfectly fit an individual user. Both types have personal value for their users but not for others.

Consider tableware for everyday use. In Japan, the teacups, rice bowls, and chopsticks are often customized. The father has his own rice bowl, the daughter her own chopsticks, the grandmother her own teacup. Japanese people tend to grow attached to tableware that they have been using for years. Their favorite chopsticks have worn edges, the teacup has a chip, yet still we don't throw it away.

In the West, spoons, plates, and glasses are all shared by the family as if they were in a restaurant.

The chopsticks and bowls don't necessarily lose quality after long use. Just as a fountain pen that has been used for years has finally been rounded in just the right way, the well-used tableware becomes comfortable. A pen tip is worn based on an individual's writing habits, so a writer feels a sense of oneness with a well-used pen.

Through many years of use, objects like baseball mitts and silver-coated lighters start to feel customized, like a life partner. Sometimes these objects wear down, making them harder to use. But if you stick with them, you can develop a fond attachment to even fragile objects. This is a relationship that lasts through thick and thin, almost like a human relationship; the tool becomes in effect a loyal companion. This is the ultimate form of customization.

Master craftsmen have such a long, deep relationship with their tool sets that the tools almost become part

Decorations help customize and personalize a cellphone.

of their bodies. It's said that when a master craftsman lends his tools to others, their delicate customization is so affected that after their return, he can no longer use them as easily as before.

A master craftsman is almost like a cyborg because his tools have become an extension of himself.

Cellphone decorations evolve

While baseball mitts and fountain pens can become highly customized, what about high-tech gadgets like the cellphone?

Cellphone straps are very common in Japan, but most cellphones in the West lack strap holes to this day. What gives?

The Japanese don't favor cellphone straps for some utilitarian reason. They are primarily accessories, a way to personalize the phone. It's akin to a little girl getting a new handbag for her Barbie doll.

Young people in Japan like cellphones with removable, customizable front casings. Returning to the doll analogy, it's like changing the dress on your favorite Barbie. Children everywhere can relate, and so, it seems, can young adults in Japan. Cellphone straps and similar products sell well in Japan because the Japanese are still connected to their childlike spirit—a connection that many Westerners would mock.

The latest rage among high-school girls is decorated cellphones, adorned with original patterns of tiny glass beads and tiles. This is a big step in the evolution of cellphones from simply utilitarian devices to highly personalized items.

A similar concept is also spreading on computers in the form of text messages with personalized backgrounds. This is called *decomail* in Japanese. More and more users are decorating the backgrounds of their messages the way they do the cases of their cellphones.

Where did all these bullet trains come from?

Visitors to Japan are often astounded at the sheer variety of goods available here. It's not only the variety of high-tech gadgets that causes jaws to drop, but the varieties of stationery items, toys, cosmetics, and toiletries too. Often Japanese themselves are unaware of the abundance of offerings they are blessed with until a foreign friend expresses amazement at all the women's hair products or ready-made food available in department stores.

Drop by a department store, head to the pickle (*tsukemono*) section, and see for yourself. A wide array of these pickles can be found, from the most popular types to specialties from a particular region. When the department store holds a pickle fair in its event space, even more varieties of pickles are on display. The Japanese consumer expects no less.

This expansion of product variety is closely related to the process of customization. Through trial and error, a manufacturer adds new varieties to its product lineup in order to stand out from the competition and appeal to highly selective consumers. It customizes its products as much as possible to individual consumer tastes.

One interesting instance of this phenomenon is the Japanese bullet train, or *shinkansen*. France's TGV, a high-

Japan has many kinds of bullet trains, and some almost appear to have been given the -tan treatment.

speed express train often considered a rival to Japan's bullet train, comes in only a few shapes and sizes. The bullet train comes in many varieties.

In 1987, Japan National Railways (JNR) was privatized and split into six companies. A quarter of a century later, these companies now operate fifteen types of bullet trains. In the West, by contrast, each country usually has only one type of super express train: Germany's ICE, Spain's Targo, Italy's Pendrino, Sweden's X2000, and the Acera of the United States.

Practically speaking, there is no need for a wide variety of bullet trains. Each Japan Railways company uses the same set of specifications. The only thing they need to change is the paint job.

But the Japanese feel that if a bullet train runs

through their town, it should be different from the "original" (that is, the Tokyo-based) *shinkansen* and have its own features. This is an example of how a culture of fine-tuned customization has developed in Japan.

Masters of the universe

Why are there so many different products in Japan? Two reasons come to mind.

One is that Japanese consumers have a strong desire for something a little different or special. This allows products with minute differences to find a foothold in the marketplace.

MBA courses in Western business schools often teach that products able to satisfy Japanese consumers can succeed anywhere. Japanese are very selective and look for a variety of options, making it hard for Japanese manufacturers to slide by with a one-size-fits-all business plan.

If a maker believes, for example, that consumers will be satisfied with a cheap detergent as long as it does a fairly decent job of washing clothes, it can never create products that appeal to picky Japanese consumers.

Another reason for the wide variety of products is

that Japanese manufacturers have a deep sense of vocation and dedication, with the ultimate aim being total mastery of their field. In Japan, we call this the path or the way (*do* in Japanese).

Originally, the concept of a "way" implied rigorous training in such traditional disciplines as *kendo*, the tea ceremony, and *kado*, or flower arranging. But nowadays every job or avocation has its own "way," including *yakyudo* (the "way" of baseball), *yosetsudo* (the "way" of welding), *henshudo* (the "way" of editing), *eigyodo* (the "way" of sales), *sommelier-do* (the "way" of wine tasting), and *karaoke-do* (the "way" of karaoke).

Long before the word "*otaku*" was coined, Japanese railroad freaks studied not only the thirteen types of bullet trains, but the many other express trains in Japan. They mastered "*tetsudodo*"—the "way" of the railroad.

In pursuing a "way," Japanese typically move beyond an interest in craftsmanship to a kind of sacred search for the ultimate. This mindset encourages stoicism and, in extreme cases, the seeker ends up ignoring the commercial side of the business and becomes obsessed with creating the ultimate product.

The dynamic between the dedicated maker pursuing its "way" and the picky consumer seeking distinctive products creates a multiplier effect, as the former tries to satisfy the latter to the maximum extent possible. That's why Japan has thirteen kinds of bullet trains.

Selective customization with home electronics

A similar sensibility exists among Japanese audiovisual makers, who incorporate customization such as color adjustment for TVs and sound equalization for stereos into their products. Users can fine-tune audiovisual output to their exact preferences.

Electronic appliance makers go even further in designing products to satisfy picky consumers, from televisions and stereos to household appliances. The analog (i.e., nondigital) product in the home that best exemplifies this extreme attention to consumer needs is probably the massage chair. Real Pro, a massage chair made by Matsushita, changes its program in the course of a massage even in the automatic mode. When the user adjusts the position of the rollers or repeats certain movements, the machine incorporates these changes into its program, thus evolving in accordance with the user's preferences. Each time the user enjoys a session with Real Pro, the program comes closer to that user's personal massage ideal.

Japan's massage chairs can remember patterns of use and adjust in response.

Even household appliances with simple energy-saving features have undergone the customization process. Odoridaki, an electric rice cooker made by Sanyo, offers more than 100 settings designed to accommodate specific strains of rice and the needs of the user. The taste of the cooked rice differs depending on the strain, even if the same setting is used. Also, the user can choose from 125 options to suit individual preferences. The user can even choose the flavor of the rice itself.

Washing machines also have advanced customization functions. Himawari, a fully automatic washing machine made by Sanyo, allows the user to set the time for the rinse and spin cycles to the exact minute. In the days of the dual-drum washing machine, which required the user to manually transfer laundry from one drum to another between cycles, housewives had to come up with their own time periods for washing, spinning, and rinsing. Himawari's precise time settings help people doing the laundry waste less time.

This preoccupation with customization is often brushed aside as sentimental nonsense. But a desire for individualized and fine-tuned customization is naturally born and bred in the Japanese.

Massage chairs are not the only Japanese machines able to remember patterns of use and then adjust in response. Computer makers have similarly customized their products to the needs of individual users.

When you write Japanese with a word processor you need a *kanji* (Chinese character) conversion function. When you type a word in *kana* (Japanese syllabary), the computer gives you corresponding choices in *kanji*. Japanese is written with a combination of *kana*, *kanji*, and occasional roman letters, but excessive use of *kana* is considered childish and unsophisticated. Also, *kanji* clarify meaning in a language with many homophones.

Modern computers also remember your conversion pattern and display high-use *kanji* first. So when I use someone else's PC, there's something strangely different about the way it works. Also, since it knows its regular user's conversion pattern, I feel as though I am being given a glimpse inside that person's mind. For example, if I try to input "I want to marry you" (*kekkon shitai*) and get the *kanji* for "bloody corpse" (also *kekkon shitai*) I'd wonder what that person was thinking.

Making *kanji* appear in their order of use is the first step to customizing the Japanese computer. But there are more to come.

You've probably heard the term "personalized search." As the Google search engine repeats searches for information on the Web, it learns the user's search pattern. In Japan, we have a saying, "十人十色," which means roughly that when ten people hear something, they call up ten different images, or that there are as many opinions as there are people. For example, when Engineer A hears the word "silver" (白金), he may think of a catalyst used in a catalytic convertor. Meanwhile, Mr. B of the Financial Affairs Department may think of precious metal prices in the futures market, while Miss C, a college student, may imagine a skin cream containing trendy silver flakes. PCs will advance from a personalized search function to the point where, after you input "silver" on the keyboard, you hit a "personal button" or a "me button" and search results will appear that follow your thought pattern. The more you use this button, the more it becomes a smart partner that almost seems to read your mind.

Also, when you get stuck for ideas, you can press a "me two years ago" button or a "department manager Yamada" button sourced from a knowledgeable superior in your office. You could also push the button of Nobel Prize winner Koichi Tanaka, writer Haruki Murakami, baseball superstar Ichiro, or other people whose opinion you respect.

This is somewhat different from the approach of conventional computers, which have long been considered devices that seek only one correct answer. (The earliest computers were essentially calculating machines that crunched numbers to get right answers to mathematical problems.) But the concept of a "me (or someone else) button" pays respect to each person's individuality, while adapting to the user's value system.

This, I believe, is the logical path of computer progress. The aim is a computer not limited to one correct answer and not thinking in place of its human user. Instead it can be an advisor that implicitly understands its user's mind.

Individualized products on flexible assembly lines

I mentioned the changeable cellphone, where users decorate their phones with beads and tiles. Take this desire to individualize a step further, and you can imagine a time when we actually assemble our cellphones to suit our personal tastes.

While the type of person who buys electronic parts at Akihabara and assembles his or her own cellphone hasn't evolved yet, plenty of computer geeks are already doing something similar with their computers. The cellphone geek can't be far behind. And once those geeks appear, they are going to be far more attached to their self-made cellphones than we are.

Mass-produced objects such as PCs become handmade objects once the computer geeks customize them. No two are alike. As the demand for customization grows, would it be possible to change the mass-production method, which pays careful attention to achieving uniform quality? Is the opposite of the current system possible?

I believe it is possible. And the time is coming when there will be a strong demand for it. Imagine an approach that avoids uniformity and allows for variations as long as the core functions are not impacted. I call it the "flexible assembly line" approach, and its time is almost here.

In the era of the flexible assembly line, you and a friend may buy the same cellphone, but when you shake your cellphone, you hear parts rattling around inside. Meanwhile, your friend has three secret buttons on her phone and part of the chrome is missing from the antenna. That's the kind of era it will be.

Rattling parts and missing chrome result in customer complaints today. But in the future users will feel affection for such defects, thinking that they express individuality. Knowing this, makers will deliberately produce such imperfections. This new "customized" mass production era is right around the corner.

Create Compulsion:
The Art of Touch

Removing the barrier between humans and machines

Tools and products are designed with human physiology and psychology in mind, but they don't necessarily move the same way humans do. Some products move in ways that are humanlike—but inside, completely different principles are at work.

Automobiles, for instance, move by the rotation of tires, which is different from human locomotion. The rotating motion of tires, which functions best on flat surfaces, has rarely emerged in 3.5 billion years of biological evolution, the few exceptions being microorganisms with flagella and other screwlike locomotion mechanisms.

The field of data processing is the same. Human brains and computers recognize images using completely different methods. For example, computers find it unexpectedly difficult to pick out images of a white dog facing right from many similar images. And if you add more ambiguous terms such as "young mongrel" or "frolicking dog," the computer is at a complete loss. Though it is easy for humans to recognize these images, a computer cannot because its recognition methodology is fundamentally different.

As indicated by these examples, the basic principles of human and machine movement are different, which acts as a barrier to interaction between the two. This barrier is integral to the human-machine interface.

Breaking through this barrier is a major challenge for product developers. In recent years, the focus has been on concepts of "barrier-free" and "universal" design, which makes products that the elderly and disabled can more easily use. Design that allows the average person to use something quickly without the need for a thick manual is called "affordable design." The user intuitively understands how to use the product on sight, thus immediately breaking through the human-machine interface barrier.

A good example of affordable design is Japan Railway's Suica automatic ticket wicket. Designed by Shunji

Yamanaka, an expert in human engineering, the Suica card reader has proven to be problem-free and popular ever since its introduction. The beauty of the Suica system is that just by looking at the card, the commuter easily understands that he or she must hold the ticket over the gate's card reader to open the ticket gate.

Engineers strive to develop technology that lowers the barrier between humans and machines, reducing human stress to as close to zero as possible. The next step after eliminating stress is design that makes people want to use the product.

In essence, the aim is make consumers instinctively want to touch the product and try it. And once they try it, they should want to keep using it, even become a slave to it. If the maker accomplishes any of these goals, he or she will be more than happy.

A self-destruct button that doesn't detonate

Zarigani Works is a company established by two artists who design and develop toys and so-called character goods. One of their products is a "self-destruct button," a push button toy inspired by a standard feature in adventure anime for boys. It's a toy version of the "final button" that the hero or villain presses in desperation, like the ejection button in a jet-fighter cockpit.

Insert the key, turn on the electrical power, open the safety cover, and then push the button.

But nothing happens . . .

To Zarigani, that's exactly the point. If a bang or flash occurs, the product has been reduced to an "activation button" that causes something to happen. Since nothing happens, the user can imagine his or her own scenario in which something does occur. The self-destruct button operates according to the Japanese aesthetic principle of *wabi-sabi*, which might be roughly defined as "less is more."

In essence, switches and buttons are used to start an action. But the whole point of this suicide button is simply to be pushed. The concept, you might say, is to give the button, not the action it incites, ultimate pride of place.

When given an ample budget and free rein, engineers tend to prioritize quantification, which means to express effectiveness with numbers. Such expressions as "doubling operational features" or "reducing response speed by half" indicate an increase in the quantity of features. "Prioritizing quantification" means to incorporate

as many features as possible into a product while minimizing the user's perception of complexity in the product and its component parts.

One example of this quantification process is the cellphone. Instead of simply being a means of voice communication, cellphones today offer supplementary features such as playing music and conducting financial transactions. By tapping the glass of the touchscreen, the user can easily access any number of applications, with almost no sense of operating buttons or switches.

Designer Naoto Fukasawa took a very different approach when he designed the popular Infobar model for the AU cellphone service launched by KDDI in 2003. The raised keys of the Infobar are transparent tiles, designed so the user cannot resist touching them. When the folding cellphone was the norm, the Infobar phone, which doesn't fold, became the stand-out in the marketplace. The Talby phone designed by Mark Newson and the egg-shaped Penck phone designed by Makoto Saito, with its glossy mirror surface, also embody the concept that buttons have a curious attraction: People want to touch them.

The buttons on these phones, unlike the "invisible" cellphone buttons mentioned above, proclaim their existence and compel users to touch them. It's the same concept as the suicide button. Seeing these buttons, users instinctively understand that their entire purpose is to be pressed. The Infobar, Talby, and Penck cellphones skillfully exploit this psychology.

I've got to press that!

The suicide switch expresses the idea that the forbidden is somehow compelling. Everyone has felt an impulse to pull the fire alarm or touch the wall with the "wet paint" sign, though they rightly seldom act on it. Children, who are naturally curious, find it difficult to repress such impulses. Let's analyze this psychology.

So many small pleasures

Various sensual situations compel people to act impulsively, but the strongest are ones associated with touch. The sensation of different textures on the skin or the tongue, of chewing food, of clothing on the body, of sitting on a comfortable chair—all are intrinsically appealing. Call them small pleasures. As a result, touching becomes habitual.

Many product features attract attention because they are visually pleasing or easy to use, but it is the act of actually touching them that qualifies them as a small pleasure. In many cases touch is the primary attraction. These products stimulate users' pleasure centers, but I'm not sure how consciously their designers have incorporated the "compulsion" element into them. In any case, in designing products it's important to be conscious of such an element.

I have compiled a list of things that people touch or manipulate impulsively in their everyday lives and categorized them as follows:

COMPULSION TO TOUCH
- The pleasure of fine, smooth surfaces (hair, silk, pantyhose)

- The pleasure of soft, smooth surfaces (velvet, suede fabric)
- The pleasure of soft, gel-type objects (cat paw pads, succulents)
- The pleasure of raised surfaces (raised seals, pens with ribbed grips, embossed paper)

COMPULSION TO PEEL

- The pleasure of scratching something off (Scratch Lottery tickets)
- The pleasure of peeling off (facial packs, adhesive duster paper)
- The pleasure of stripping off (protective sheets for liquid crystal displays)

Similar pleasurable actions include extracting crab-meat from shells, peeling off scabs, and scraping off the burned food stuck to the surface of a fry pan.

COMPULSION TO VIOLATE RULES

- The pleasure of doing something forbidden (pulling a fire alarm)
- The pleasure of violating a written or verbal injunction (touching a "Wet Paint" sign)

COMPULSION TO CRUSH

- The pleasure of pressing something completely flat, such as an empty can
- The pleasure of popping sheets of bubble wrap
- The pleasure of grinding sesame seeds, coffee beans, or rock salt in a handmill
- The pleasure of tearing paper, string cheese, or faux crab sticks; breaking apart disposable chopsticks; removing the laminate film from packaging

Similar pleasurable actions include stepping on the morning frost or crushing fish eggs between your teeth.

Inexplicable pleasures

A good example of a slightly different type of product that people like to touch is Clip-on, a multicolor ballpoint pen made by the Zebra stationery company.

Multicolor pens have been around for years. What's distinctive about Clip-On is that it is equipped with a mechanical reset switch in addition to the normal slide mechanism for changing colors. Moreover, the clip that attaches to the breast pocket has a hinge mechanism that opens and closes like a lever.

It's hard to keep from clicking away with the Zebra Clip-On.

Zebra calls the slide, push, and hinge features the "three big switching mechanisms." These mechanisms are installed on a part of Clip-On in easy reach of the thumb of the hand holding the pen. Clicking this pen takes on a different meaning compared with other multicolor pens.

One can easily imagine a supervisor absent-mindedly fondling his Clip-on pen during a meeting. Repetitive, unconscious movement of the fingers is said to relieve stress, just like chewing gum. And supervisors tend to build up a lot of stress (though they are not the only working people who fondle pens).

This particular pen is a standout for its superb man/machine interface; it is truly a friend to your fingers that keeps them entertained and occupied while the brain is working.

Create Compulsion: The Art of Touch

Pens may someday incorporate such features as audible musical notes corresponding to the five colors or some kind of rotating mechanism. Such features would allow the supervisor more options for relieving stress and perhaps help him come up with better ideas (though he may also end up annoying the people around him).

Normally, pen R&D involves adjusting the colors or improving the ballpoint so the ink doesn't smear—both never-ending technical challenges. But it is also important to understand other factors contributing to consumer appeal, such as a compulsion to touch and a desire for small pleasures, and design them into products that offer cutting-edge technologies. These considerations are especially important in making products for the Japanese, who tend to become attached to their tools.

It isn't easy to incorporate a compulsion to touch into a product. Engineers prefer functionality because functional effectiveness can be expressed in numbers. If you allow engineers to do as they please, the pen of the future will incorporate all sorts of electronic technologies whose effectiveness can be measured quantitatively.

Of course, a pen with many versatile features is not necessarily a bad thing. At the 2004 CeBit electronics exhibition, the German company Siemens unveiled the prototype of such a pen: the GSM Terminal Pen Phone. This pen is equipped with a cellphone compatible with GSM systems (second-generation cellphone systems used in many countries at the time). When users wrote numerals on the phone, it could recognize them as a telephone number to call. When users wrote text, the phone could recognize it as a message to send. It also has a voice-recognition feature and could connect to a personal digital assistant by means of a built-in Bluetooth headset. It's like something James Bond would use.

Technological advances will also come from human engineering, an academic discipline that studies tools or products from the viewpoint of ease of use, based on human physiological and psychological characteristics. Engineers take the structure of human fingers into consideration when designing pens. They'll research shapes that don't cause hand fatigue even after long use, for example. This field of human engineering is called ergonomics. One successful example of ergonomics in action is the Ergo Pen, a writing instrument whose grip is fashioned from rubber-based materials molded into a gentle curve.

Ergonomics is a recognized academic discipline that can measure results quantitatively and thus achieve clearly defined research targets. Among the problems it

The Ergo Pen is perfectly balanced to reduce fatigue and muscle strain.

addresses are improving the productivity of assembly line workers and reducing the fatigue caused by operating airplanes or automobiles for long periods.

Ergonomics has more serious purposes than designing pens that are enjoyable to fondle. Ergonomists would balk at analyzing how to turn a pleasurable action (fondling) into a compulsion in order to create appealing products. The reason: pleasure and compulsion are difficult to evaluate quantitatively, unlike production efficiency or degree of fatigue.

How pen twirling became a global fad

Let's look at another way pens give tactile pleasure: pen-twirling. This form of recreation emerged in the mid-1970s and was popularized by Japanese cram school students. Masters of the basic technique, which involves dexterously spinning the pen around the top of the thumb, can twirl a pen reflexively while thinking about other things. In the 1990s, pen twirling became a phenomenon among young people worldwide thanks to the Internet. However, Asians, in particular Japanese and Koreans, seem to partake of this pastime more enthusiastically than others.

There are even competitions for pen-twirlers. Techniques are graded by degree of difficulty and include "normal," "reverse," "single axel," and "sonic." Hundreds of techniques have been created, some of which are quite difficult.

Enthusiasts exchange bits of information, such as, "Pilot's Dr. Grip has good heft, so I recommend it for beginners," or "Sunstar's Twin Marker is good for doing sonics." They study this information to choose the right pen for their twirling needs. This phenomenon is a business opportunity for engineers versed in such matters as center of gravity, moment of rotation, coefficient of surface friction, and the appropriate hardness for objects in motion. Since a demand exists, engineers can design a pen that's easier to twirl, knowing it will sell. They can also adjust the design for a specific technique or make exchangeable parts so the user can customize the pen to fit his or her twirling style.

Pen twirling, which has the characteristics of a touching compulsion, is quite addictive. Such an addiction presents a cornucopia of business opportunities.

Other compulsive actions, such as leg bouncing, nail biting, and hair twirling, also relieve stress. Among compulsions using objects at hand, one of the most common is fiddling with the receiver cord while talking on the telephone. Sometimes when I am in a conference, I imagine telephone cords dangling from the ceiling to each seat so that participants have something to fondle. It might improve the effectiveness of meetings.

Products born from compulsion

Touching, as I've mentioned, can become a compulsion through such acts as habitually crushing bubble wrap, fondling the paw pads of cats, and clicking ballpoint pens. One can easily imagine a child becoming obsessed with such actions and doing them all day long. I'm sure many parents have seen their toddlers compulsively empty the contents of tissue boxes.

But when a grownup loses the childlike qualities

associated with these kinds of compulsive actions, he falls into a debased state of adulthood and can no longer recognize the value and significance of his compulsions. They become simply mechanical.

Many Japanese adults, however, retain a childlike love of compulsive actions. For companies, this offers certain Japan-specific business opportunities.

One example of a successful product that cleverly exploits the compulsion concept is the massage chair. It's more accurate, though, to describe it as the object of a pleasurable obsession rather than a mechanical compulsion. In any case, the massage chair is a standout among machines that give users pleasure in their own homes.

The ancestor of the home-use massage chair was the coin-operated massage chair traditionally found in hot-spring resorts. The coin-operated massage chair was an unsophisticated device: Users had to turn a knob on the side of the chair to manually manipulate the rotating massage balls.

First marketed globally by Matsushita Electric in 1969, the home massage chair has been evolving ever since. The company's most advanced model is the Real Pro series, which massages the entire body, including the soles of the feet, the calves, the thighs, the lower back, the upper back, the arms, and the neck. It features a drive mechanism with various motors, air bags, and air pumps controlled by computers that can perform nearly one thousand separate massage actions. An applied specific integrated circuit (ASIC) system built into the chair controls these elaborate movements.

The chair, which massages the user's entire body, looks like something from the future. It's like a robot-controlled seat in which a human is encased. All Japanese are familiar with a children's song written by Saijo Yaso that contains the lyric "Mother, let me pound your shoulders, *tanton, tanton, tanton-to.*" (*"Tanton"* mimics the sound of pounding.)

Japan's culture of home massage, which creates a compulsive need for the pleasure derived from touch, remains alive and well. This massage tradition has contributed to the popularity of the home massage chair.

Another Japanese custom associated with pleasant feelings is ear cleaning.

A common activity in Japan, with ear picks available in all train station kiosks and airport souvenir shops, ear cleaning has strong nostalgic associations since Japanese mothers often clean their children's ears. Also, some grown men with traditional ideas about their preroga-

Japanese mothers often clean their children's ears with picks like this bamboo one with a cotton puff.

tives as kings of the castle enjoy having their wife clean their ears on their days off. The man will typically lay his head on his wife's lap as she cleans his ears.

To Westerners the custom of ear cleaning may seem peculiar. This is not a family activity, after all. Some may go to an ear-nose-throat specialist to have their ears cleaned, but few Western men would have the temerity to ask their wives to clean their ears!

Japanese barbers customarily give their customers neck massages and clean their ears following a haircut. Unfortunately, they usually finish their ministrations just when the customer is starting to get into it.

According to Ueno Rei, a professional ear cleaner and expert on the subject, the custom of ear cleaning spread among the working class in the Edo Period

(1603–1868). During the time known as the Kansei Era (1789–1801), barbers called *narabidoko* lined the streets in the vicinity of the Ryokoku district in the city of Edo (present-day Tokyo). These barbers not only cut the hair but cleaned the ears of their customers, who spread the word about how good it felt.

The Japanese have devised a wide variety of ear-cleaning devices. In addition to the conventional spatulate ear cleaner, there is also the narrower *nonoji* type, characterized by its thin ringlike shape, and a coil type that looks like an expanded spring. There is also a suction ear cleaner that gives a view of the inner ear through a loupe, as well as one that removes dry skin using an adhesive. The mechanics of ear cleaning have been perfected in a very Japanese way.

Someday an ear-cleaning device may appear with built-in mechatronics, a field in which Japan excels. The desire to perfect the Japanese way of ear pleasure is insatiable and serves as an impetus for creating state-of-the-art products.

Leave *Something* Out: *Sundome*

Can a product be too convenient?

First, let's suppose that technicians, after much painstaking effort, have developed tools and products with absolutely no interface barriers. This is only a supposition, since it would actually be impossible to do. But would using such tools and products be enjoyable?

Let's take musical instruments as an example. Let's say there is a piano that allows you to easily play exactly what the composer has in mind. You do not have to undergo the difficult process of actually learning how to play.

Many people learned the piano as children, but have since given it up. But for this piano, which anyone can play, hard practice is not necessary. Anyone can skillfully play any piece of music on it, no matter how difficult, from the day they buy it.

Some may think that such a piano, on which anyone could perform their favorite tunes like a pro, would be a hot seller. But I disagree.

This instrument may increase the number of composers, but it will not be as popular as ordinary instruments. Why? Hard practice and polishing one's talent, as well as interpreting a piece of music in one's own way, have an intrinsic meaning and enjoyment.

Through the process of polishing his or her technique, a pianist can produce tone qualities he or she had never thought of before. Even performers who play without improvising can astonish listeners, who can't believe the sounds they are hearing. They praise the talent and effort of the performer who can play in such a superhumanly skillful way.

But a piano that allows one to easily play any difficult piece of music would fundamentally undermine the significance of the act.

I would like to propose "stopping short," or in Japanese, *sundome* (soon-doh-meh), as an important concept for future manufacturing. Instead of pursuing perfect ease of use as the ultimate good, this concept exults the effort needed to use a tool or product effectively and the "joy of use" experienced as compensation for that effort.

The martial arts rule that one should not touch an opponent's body with a kick or punch is called *sundome* (literally, "stopping short"). Manufacturers should make tools and products that "stop short" of zero effort. If they do not, the enjoyable labor required to master a tool, through a process of trial and error, devolves into a simple, boringly easy operation.

Enjoying the process of mastery

Among my acquaintances is an elderly woman named Miss Kobayashi who cleans buildings. For 40 years, starting from about 1945, she worked as a typesetter. From the end of the Meiji Period (1868–1912) until the 1980s, when word processors came into general use, workers in Japanese print shops had to set type one character at a time. (Japanese uses not only Chinese characters, but also the native syllabaries *hiragana* for Japanese words and *katakana* for foreign words, as well as the Roman alphabet. For the sake of simplicity, however, I will refer to all as "characters.") Type was made of cast metal and looked like seals used for affixing signatures. Once the type was set, the manuscript could then be printed.

A fast worker, Miss Kobayashi could set the type for one newspaper page in one day. It would take her about one hour to set the type for one "sheet" consisting of 25 characters to a line and 53 lines. Given that there were eight working hours in a day, she could complete eight sheets a day. That would be the equivalent to about one page of a newspaper.

In other words, she was setting one character every two or three seconds. This was a fast pace even among her workmates at the print shop. Since an amateur would need nearly ten minutes to set one character, we can see how extraordinarily fast Kobayashi was.

Kobayashi and other experienced typesetters could chat with the people next to them as they worked. The characters, she said, seemed to jump out at her. Her fingers would set the type on their own; she didn't have to think about it. A male co-worker next to her would arrange type to read, "Let's see a movie together after work tonight" in a box next to her and she would reply, "I'm free tomorrow" with type in the same way. She had pleasant memories like these of the sophisticated tricks of her trade.

A well-trained typesetter could pick up characters by reflex, just as ordinary people can drive a car without being aware of what they are doing or a pianist can send his or her fingers flying over the keyboard without thinking. But since the end of the 1980s, use of word processors has become widespread and many typesetters have lost their jobs.

A keyboard takes far less effort than setting type. If with a little practice you obtain a third-class certification in Japanese word processing, you should be able to type one character (that is, one Chinese character) every two seconds. You can "set type" at the same pace as the experienced typesetter Kobayashi.

The so-called *uchite* ("shooters") are word processors in a master class. They can work three times faster than someone with a third-class certification. Some *uchite* can even strike four or five characters a second, or the unbelievable rate of 265 characters a minute.

Such masters, who continually polish their skills while aiming for a high level of achievement, know the *sundome* enjoyment and satisfaction of reaching the ultimate level of their craft.

In the fields of both typesetting and word processing, workers are not simply composing type but also enjoying the process of mastery. If manufacturers forget that and make a product that does not involve that process, even to a limited degree, only a few people will be interested.

When people use a tool, they usually feel some sort of tension. But as they become more practiced, they naturally acquire skills. Some become so proficient that they surpass the expectations of the tool's designers. As highly skilled users of the tool grow in number, some acquire even more astonishing techniques and achieve the exalted status of "master."

Olympic gymnastics competitions now begin with tricks that were at the E level of difficulty at the 1964 Tokyo Olympics. The eye-popping new tricks that now appear in rapid succession in gymnastics competitions have attained the ultimate G level. No one pays any attention to the E-level tricks of the past.

Voice-inputting technology may someday be used for word processing in place of a keyboard. Then inputters will emerge who are skilled at a special style of speaking that compensates for an imperfect voice-inputting device. When the level of general skill rises high enough, voice-inputting masters will appear. Some masters may even speak to their word processors in voices unlike human speech.

To sum up, the pleasure people derive from the

process of mastery is connected to technical advances and the development of new products. Advances produce new challenges—and offer new pleasures from mastering them. But if technology progresses to the point where no mastery is necessary, pleasure also disappears. *Sundome* means to stop short of that point.▯

Leave room for people to display their talents

We can say the same thing about sporting goods and video games.

For example, golfers compete to see how well they can hit the ball with these awkward things called clubs. Of course, they are enjoying the process of rigorous practice.

With video games, the key to success is making hits quickly with the controller to run up points, as seen in the career of Meijin Takahashi, a famous game master with a super trigger finger. Not just anyone can achieve such speed, but striving for it is part of the fun. A game in which any player, no matter how clumsy and slow, could get a perfect score from day one would never sell.

A player who can skillfully use tools made deliberately hard to master, while following the rules, is praised by others. No one would want to see baseball played with bats that look like tennis rackets and always hit home runs. Human beings admire those who have achieved mastery, not those who have everything easy.

Elite athletes display their talents by striving for goals deliberately set at high levels using methods made deliberately difficult. By displaying their skills this way, they can be respected as professionals and earn a lot of money. But how can ordinary people display their talents in their daily lives? In searching for the answer, we can discover hints for future manufacturing.

Everyday life holds many opportunities to display talents that can raise one's own stock in the eyes of others. Examples include the ability to skillfully sign the guest book at a friend's wedding; fluently draw a caricature of a bar hostess on the back of a coaster; make book shelves in just half a day at the request of your wife; surprise everyone by playing the sax at a party; or demonstrate slick knife handling skills at a barbeque. Even an elderly man who types with blinding speed the first time he touches a personal computer can impress others.

For the performer, this is a moment when he can shine. Especially in this day and age, when so many people have to sell themselves as personal "brands," this kind of performance can be a plus.

To repeat, human beings are looking for, not perfect ease, but rather a *sundome* experience in which they can test themselves while tasting the joy of mastery, even in everyday activities. In developing products and technologies, we have to make *sundome* a priority.

The preceding examples of reputation-boosting activities can be divided into two categories: performances, such as sports and music, and practical activities. Sports and music often involve performing for an audience. Carpentry and knife handling, on the other hand, are activities that are practical for everyday life. Essentially they are not performances for others.

Some would argue that the *sundome* concept is inappropriate in designing tools for practical use. The more convenient the better, they say. Discard this *sundome* idea and strive for full automaton that makes operation easy for everyone. But I disagree.

How far should we take automation in daily life? The answer is a matter of degree; it depends on individual feelings of satisfaction.

Car-navigation systems steal the spotlight from men

Many drivers complain about the hassle of parallel parking. But parallel parking can produce an unexpected benefit: Women are attracted to a man the moment they see he can easily parallel park his car.

A man who twists his body sideways and, looking over his shoulder, smoothly parallel parks a car looks dependable to women. Even though this may not equal the skill of F-1 drivers, who can sense the width of a car in units of millimeters, the ability to parallel park allows a man to demonstrate his fine spatial sense and car-handling skills.

Parallel parking, however, is now close to being automated, and the day will soon come when anyone can parallel park automatically just by pressing a button.

For a man, actions such as parallel parking are important when he is trying to impress a woman. Good parallel parking demonstrates the superiority of his genes, much like a male pheasant spreading his wings or a male cicada calling loudly.

That's why men are tired of getting shown up by their car-navigation systems.

Do you look at a map or ask someone for directions

when you are out driving and not sure of the way? Male and female drivers answer this question quite differently. In general, men are good at finding their way by looking at a map, as though standing on a hill and taking a commanding view of the countryside.

When a dating couple goes on a drive, the man is usually the one behind the wheel, as has been the case for generations. Until car navigation systems became popular, the question of how to reach the destination without mishap was a crucial one for many men. They could not bear the humiliation of getting lost and needing to ask someone the way with a woman sitting beside them in the passenger seat.

In contrast to men, women drivers tend to readily ask the way at convenience stores or from passersby on the street. And yet, they don't refer to maps as much as men do.

Car-navigation systems are a welcome technology for guys who are incompetent drivers. But for their more competent brothers, arriving at the destination without car navigation or asking for directions is reputation-bolstering behavior that, like parallel parking, allows them to enjoy a *sundome* experience of mastery. Car navigation, in which this sort of experience no longer exists, deprives these competent guys of a manly moment.

The same scenario is also true in shopping. In general, when men enter a store, they don't ask the clerk where the shoes are. Instead they look for the shoe section themselves.

Asking the clerk may be easier, but by searching on their own, they can have a *sundome* sense of accomplishment of finding something after a search. This is reminiscent of the pleasure experienced by ancient hunters of tracking down game.

One company in Japan has skillfully exploited this feeling: the Don Quijote discount store chain. It offers customers the pleasure of finding the product they are looking for from among the jungle-like chaos of items on display, in something like a treasure hunt.

Don Quijote's method of displaying goods would also work in a Southeast Asian open-air market or a Middle Eastern bazaar. Taking the opposite approach from the Western style of tidy, easy-to-understand displays, Don Quijote offers a *sundome* experience to its shoppers.

The scientific approach to merchandising says stores should enable customers to easily, quickly, and efficiently

Shopping at a Don Quijote store in Japan is like going on a treasure hunt.

find what they want. By aiming solely for convenience, however, their displays become boring, and shopping, which should be enjoyable, becomes a simple task of stocking up on supplies.

The same thing has happened to types of housework that should be enjoyable, such as cooking a meal. What does it mean to fully automate the act of cooking? You end up warming up a pouch of curry and a pack of rice in the microwave, pouring the former over the latter and eating it.

Or to take it one step farther, you also buy precut vegetables sold at the supermarket.

But is it OK to eat that way every day? The food doesn't taste particularly good and isn't fun to make.

People want more than efficiency in their everyday

Leave Something Out: Sundome

lives. They enjoy spending the time and effort to make something tasty.

Fortunately, food companies have developed a variety of semi-instant food products for people who don't like packs and pouches, but are too busy to cook from scratch. One common example is solid roux used for making curry. After frying or boiling vegetables and meat, you only need to stir in the solid roux to make delicious curry. The reason why solid roux has maintained its sales lead over frozen, canned, and pouch curry for so many years, is that cooks can enjoy a process of cooking that is easy— but not too easy. This is yet another illustration of the *sundome* concept's appeal.

The making of curry from spices was originally a laborious task. This would not be acceptable to today's Japanese housewife. On the other hand, making pouch curry every day turns meals into a simple task of filling an empty stomach. The makers of solid roux have found a happy medium.

Direct from the brain to the machine

A car-navigation system is one of the few electrical appliances permitted to talk to humans. But soon humans may

This car navigation system is based on the popular One Piece *anime. The system speaks in the characters' voices.*

be talking to their cars. Speech inputting is already being used in today's cellphones. There will also be a shift, some predict, from pushing buttons to speaking directly to car-navigation systems.

Drivers need to keep their hands on the steering wheel and their eyes on the road, which is why car-navigation systems will probably become the most advanced consumer product for speech output and input. It just makes sense.

We'll move from manual manipulation of buttons to voice commands, but we won't stop there. The next stage will be direct input from our brains to our electronic devices. While this may sound extreme, it's already beginning to happen in the field of medicine. At Brown University in the U.S., researchers have directly connected electrodes to the cranial nerves of patients paralyzed by brain injury or disease, enabling them to move robot hands simply by thought.

This is a truly groundbreaking technology for the handicapped, but what is its significance for the development and manufacturing of devices?

Just as cooking becomes insipid and dull when made too easy, operating devices with commands from the brain would be more convenient, but not very enjoyable.

In the near future, drivers may be able to drive their cars to their destinations and park them just by thinking. As a result, experts may appear who can skillfully operate direct-brain systems. But they will not have the joy of impressing the woman sitting next to them with their parallel parking.

The makers of tools and products may be able to use the power of technology to develop direct-brain systems, but they should not underestimate the importance of *sundome*. Even if these systems become a reality, the manufacturers that will reap the rewards will be those who understand human psychology, including why people are captivated by the seemingly old-fashioned process of mastery, and use that understanding to make products. The demand for *sundome* experiences, which are not fully automated (that is, operated directly from the brain) and not completely manual either, will stubbornly survive.

Sundome luxury

The Toyota luxury car Lexus is famous for having an extraordinarily quiet interior. It's so quiet that someone with a competing German automaker once joked, half spitefully, that "a car should make a few carlike sounds." The quietness of the Lexus is so close to perfect that it's hard to know if the engine is running.

Toyota eventually came around to that person's way of thinking and decided to make the Lexus a little less quiet. Lexus engineers cut the wind and tire noise, but left some of the engine noise. Auto writer Tachio Yonemura argues that engine noise has come to be regarded not as noise, but as a "driving sound."

Technically, it's possible to make a car perfectly silent, but this is an excellent example of how "stopping short" (*sundome*) makes for success. To carry this idea a bit further, car makers could achieve complete silence, but then pipe the owner's favorite engine noise in stereo

The original Lexus was so quiet, drivers found it unsettling.

into the car's interior. For car lovers, this would be the greatest luxury.

The Japanese communication style reflects the *sun-dome* concept. This has been pointed out to me by my wife and other foreigners who have lived in Japan a long time and understand its culture.

There is an old newsreel film showing soldiers on board a ship returning from abroad. They are about to reunite with their wives, who are standing on the dock, after an interval of years. We see a ragged husband, finally back in Japan, and his wife, who has been patiently waiting for him. They catch sight of one another in the midst of the crowd on the dock and ecstatically run toward each other.

But when they come close, with just a few paces left

until they can embrace, they stop. The wife looks steadily into her husband's eyes and bows low to him. The husband replies by nodding.

In a Hollywood movie, this scene would doubtless conclude in a storm of embraces and kisses. But my wife, who has lived in Japan for fifteen years, commented that the Japanese scene made her feel more deeply the strength of their bond to each other. Hearing this, I reflected on how the Japanese have taken the *sundome* ethos to heart and how it permeates so many aspects of Japanese life.

Though not as tragic as a couple separated by war, the situation of the bride's father at a Japanese wedding is somewhat similar. The father may feel that he wants to embrace the bride, but he resists the urge—that is, he "stops short." This is a very Japanese way of thinking; men are expected to silently endure instead of physically expressing their emotions. To a foreign observer, though, it may look as though the father and the bride would rather cry than embrace each other.

But foreigners can also understand the strengths of Japan's ethos of *sundome* that considers restraint a virtue.

If Japanese companies promote the development of

Leave Something Out: Sundome

the sort of *sundome* products and services I have been focusing on in this section, they can appeal to consumers in not only Japan, but the world at large. Everyone can understand the joy of doing something well, even if it's just to cook curry rather than pour it out of a pouch. Zero effort means zero fun.

Connect Us with Kindness: Communal Values

Don't get in the way of human relations

Compared to the rest of the world, the Japanese lead their lives without thinking much about religious codes of conduct. This does not mean that Japanese are devoid of moral codes, however. Japanese adult behavior is guided by a desire to fit comfortably into the larger scheme of things while being mindful of others' feelings. This behavioral code serves in place of religious rules of conduct. Someone once called it a religion of public decency.

The Japanese refrain from expressing their opinions forcefully. Instead they tend to observe the situation from the sidelines and try to read the general mood. They then gradually form a group consensus and strive to achieve overall harmony. Prince Shotoku's famous saying from

1,400 years ago, *"Wa o motte tattoshi to nasu"* ("Harmony is of utmost importance") still resonates today.

The Japanese definition of belonging often proves mystifying to foreigners. The distinction between *"uchi"* (inside, us) and *"soto"* (outside, them) is always strongly felt. In business situations one often hears the phrase "in *our* company (*uchi no kaisha dewa*) . . ." The word *"uchi"* has nuances that are hard to translate into other languages. When one is a member of a group, the mechanisms of *"amae"* (emotional dependence) come prominently into play. Communication turns into a *"naa naa"* relationship (characterized by mutual backscratching and a clubby fuzzing of norms), and adherence to abstract rules tends to be conveniently forgotten. This kind of relationship is not necessarily a negative one. In fact, some innovative products have been born from this distinctly Japanese communication style.

Products and tools should not create friction between people. That is the exact opposite of their intended purpose, and engineers should instinctively understand this. Yet it still happens all the time. Just think of the docile, mousy guy who gets behind the wheel of a car and becomes a demon, speeding, antagonizing other drivers, leaning on the horn and running red lights.

Or the Internet troll who may be mild mannered in person but dishes out relentless hate speech in the comments sections of blogs and chat rooms. The cloak of anonymity on the Internet emboldens the troll just as the metallic armor of the car does for Mr. Road Rage.

A good engineer should feel ashamed if his or her product or system encourages such anti-social behavior. Products and tools should not create friction between people. They should instead serve as a kind of shock absorber between people. One consequence of valuing qualitative functions above other factors in manufacturing is that the resulting products, tools, and systems tend to increase the distance between people instead of decrease it. Solving this problem is more urgent than figuring out how to increase engine horsepower or computer speed.

Thanks Tail and Post Pet spread warmth

Thanks Tail is a gizmo shaped like a dog's tail that is attached to a car roof. The user makes the Thanks Tail wave whenever he or she wishes to express gratitude to another driver who yielded or performed a similar kindness. The tail moves when the driver sends an electric signal from the car seat.

The Thanks Tail wags to express appreciation for the kindness of other drivers.

The Thanks Tail can be useful in many other situations. Say you drive a girl home after a date and she waves at you as you drive off. A wag of the Thanks Tail may be just the romantic touch to cap off the evening.

Media artist Hachiya Kazuhiko created the Thanks Tail. It's available at Japanese auto-parts shops today. Hachiya is also famous as the inventor of Post Pet software, which uploads a small animal mail carrier on your computer screen that delivers your email.

Both products were conceived to act as links and shock absorbers between people. Hachiya is a talented man who gave serious thought to products that could lessen friction between people and then went to the trouble of making them. His way of thinking reflects a very Japanese style of communication.

Putting consideration for others above comfort and ease

Let's take a glance back at the long history of the automobile. Generally speaking, cars have developed in two distinct ways. One type of car is designed to maximize the joy of driving. The enjoyment comes from the driver being able to control the vehicle as if it were an extension of his or her body. Whether the driver is pressing the accelerator to experience the exhilaration of speed, steering with experienced ease through a tricky curve or parking in a tight space with perfect spatial awareness, these cars are designed to enhance the thrill of bending a finely made machine to the driver's will. The experience of driving one of these cars is somewhat like that of a cowboy riding a trusty steed over a rugged mountain range.

The second type of car is designed for luxury. These cars are made for people who can afford to hire a chauffeur so that they can enjoy a comfortable commute without bothering to actually drive. They provide a living space like that of a private cabin on a sleeper train.

The end point of development for the first type, which focuses on the exhilaration of the driving experience, is a quick, light sports car or a mega truck with enormous horsepower. A chauffeured limousine epitomizes the second type of car.

The two types of cars together are in Japan called the "two *raku*" (the Chinese character translates as both "enjoyment and fun" and "comfortable and effortless"). The history of automobile design is characterized by the pursuit of two types of luxury: fun driving and comfortable transportation.

"Two *raku*" cars are designed to fulfill the desires of their purchasers to enjoy the experience of driving or of travel in comfort and ease. The Thanks Tail fundamentally diverges from these products because it was devised to express the driver's consideration for others. This simple toylike product has ventured into a league of its own.

The Thanks Tail represents a higher dimension of what might be called mental luxury: the driver feels less stress being stuck in traffic.

Japan's driving culture has a tradition of valuing harmony. Taxi drivers dim their headlights when they stop at an intersection to reduce glare for oncoming traffic. When one driver is given the right of way by another, he or she expresses thanks by flicking the hazard warning flashers. Drivers also signal with their hazard flashers when they encounter heavy traffic on the highway, warning cars behind them of the jam.

Although Japan has its share of road rage, drivers here also practice considerate traffic manners, with rules developed naturally over the course of many years. This may sound overly dramatic, but I believe the unspoken need of Japanese drivers to express their consideration, bubbling under the surface, led to the development of the Thanks Tail.

Car manufacturers should move with the times and stop obsessing over the two *raku*. If they try to develop a new type of car based on the Thanks Tail's example, the result will be much more sophisticated—and nicer—machines.

Machines that relax your mind

Who hasn't lost their cool, at least a little, when a driver cuts you off or you sit unmoving in a traffic jam? It's easy to see why the blood boils as drivers chug along toll roads at a glacial pace.

When we deal with an unresponsive machine—like a car in a traffic jam—our stress level shoots up. Remember the last time your computer crashed? At least cars give off warning signs like smoke from the engine or unusual vibrations. A computer can just drop dead in an instant. When that happens, our sense of futility is quickly replaced by anger. We start to hate our computer, and in extreme cases, we may begin to hate the manufacturer and its entire line of products.

Common sense solutions for mechanical machine failures do not apply to computer crashes since everything is in a black box. This is the computer's biggest problem: crashes destroy whatever affection and attachment we had for the machine and fill the void with rage.

The technology of stress reduction has produced some amazing devices. One example is a traffic light that counts down in seconds the waiting time remaining on a red light. A maker with a more playful approach could create an apologizing traffic light. One inspiration would be Japanese construction signs featuring a cartoon construction worker called *ojigi-bito* (bowing man) who apologizes to pedestrians about the inconvenience caused by ongoing construction work.

The apologizing traffic light would digitally display the bowing man on the side, apologizing until the red light turns green. The strain of living in a concrete jungle like Tokyo may be slightly eased when the bowing man says, "Just a moment more, thank you for your patience," as he bows deeper.

Discreetly calling attention to ourselves

In Japan, cyclists often ride on the sidewalk. Ringing the bike bell and weaving through a crowd can be quite embarrassing for the average Japanese. It makes them feel like an arrogant shogun, demanding the right of way. They find it hard to muster up the nerve to behave that way.

Taking such feelings into account, a new product called Toryanse (the title of a traditional children's rhyme and game) has been launched to replace the conventional bicycle bell. Users attach it to their bike and press the switch to produce a ding-dong chime sound instead of a bike bell ringing. The chime is followed by a recorded voice saying, "Excuse me, may I squeeze by?"

If you've ever been bothered by the arrogant "make way" message of ringing bicycle bells, you'll appreciate the subtle Toryanse approach. The message to pedestrians seems to be, "I humbly apologize for requesting the right of way." This product, which aims to connect cyclists and pedestrians, is permeated by the gentle personality of its designer.

Let's try to imagine a more advanced version. Since the purpose of the product is simply to tell pedestrians to make way without appearing arrogant, a recorded message is not necessary. Instead the device could emit the sound of a creaking brake. Actual braking does not always produce the creaking sound familiar to many Japanese and also has the disadvantage of stopping the bike. The solution is a synthetic creaking sound at low volume to subtly warn pedestrians that a bicycle is behind them.

Pedestrians would hear the noise and associate it with actual braking, not an annoying noise that says, "Get out of the way!" The cyclist is absolved of blame and able to ride around the pedestrian without raising anyone's stress level. Or we could reproduce the ratcheting noise made by the pedals, which is also a familiar sound that no one will find obnoxious.

The Japanese value a delicate style of communication, particularly when it comes to expressing our emotions or intentions. Sociolinguistics Professor Yasushi Haga has listed eight characteristics that define the communication style of a typical Japanese person:

1. Is reticent (tries to sense the mood of others)

2. Doesn't try to persuade (first builds personal trust)

3. Expresses concern for others (reads the atmosphere of the situation)

4. Keeps ego in check (places a high value on humility and is averse to overt self-promotion)

5. Cultivates the self (a characteristic of an agricultural people who see themselves as a field to be cultivated or regard everything as ascetic training)

6. Is humble

7. Goes with the flow (tries to maintain the status quo instead of battling the elements)

8. Trusts others to do what is needed or right

Both the Thanks Tail and Toryanse are products resembling *kaeshi-waza* (a martial arts technique in which the practitioner defends against an opponent's attack while simultaneously launching his own). They defuse aggression from outside while advancing the user's interests.

Cellphone users have developed tens of thousands of emoticons in Japan, including these for expressions of surprise.

Emoticons that read between the lines

Created from existing symbols, emoticons are pictograms that express emotions. They originated in the typographic "smiley face" (:-) said to have been born in the U.S. in the 1980s. In English-speaking countries, emoticons only number a few hundred types, but in Japan they exist in the thousands. And more are being invented as I write.

Several theories exist as to why emoticons became so widespread in such a short time. Keichi Makino, a lecturer in Kyoto Seika University's Comics Studies Department and the executive director of the Japan Society for Studies in Cartoons and Comics, finds the emoticon's origins in the older culture of manga and speech bubbles, just as manga was preceded by the culture of *kanji/kana* (when imported Chinese characters were "subtitled" with a simpler, Japanese-created *kana* syllabary to make them easily readable).

Whatever their beginnings, emoticons are the most useful tools ever for communicating between the lines. When you are feeling truly sorry for a misdeed or misstep, rather than write a long letter full of excuses, it is more effective to add **m(_ _)m** (an emoticon that looks like a person with his or her head pressed to the ground in abject apology) to communicate your contrition. This usefulness is a defining characteristic of emoticons.

In the Heian Period (794–1185) Japanese women devised the *kana* syllabary. Its benefits to Japanese culture have since been immeasurable.

Japanese is the only language in the world that mixes ideographs (such as Chinese characters) and phonograms (such as the English alphabet and *kana*). Because these two writing systems are part of their daily lives Japanese seldom notice the many advantages they receive from their combination.

About a thousand years have passed since the Heian Period, and Japan again finds itself in an era of peace and cultural maturity. The resemblance between modern-day *gal* culture, which is developing a new communication style with emoticons, and the women of the Heian Period,

NEC's Kotohana can read emotions from afar.

言花
KOTOHANA

who used *kana* to create quality works of literature, is deeply significant. In peacetime, Japanese culture's feminine aspects tend to blossom and lead the rest of society.

Machines that read emotions from voices

In Japan, so many emoticons have been created that it's reasonable to assume Japanese appreciate their convenience more than anyone else. A strong demand exists for indirect means of communication.

Here is a high-tech example: In 2006, NEC and its partners launched a device called Kotohana (Word Flower) for aiding interpersonal communication. Shaped like three differently colored flowers and containing luminescent diodes, Kotohana is installed with an "emotion-recognition engine" incorporating speech-recognition technology that listens in on people's conversations. The device does not understand content but instead measures the speaker's tone, pitch, and rhythm to identify the emotions at play. It then summarizes its findings into three factors and displays them in real time via the colors of the luminescent diodes. Yellow indicates that the speaker is feeling joy; blue, sadness; red, excitement; and green, calm.

Subtle gradations indicate changes in emotion while emotional intensity is expressed by the brightness of the Kotohana's glow. Say a man asks a woman on the phone, in a breezy tone, if she would like to go to a movie. Even if the man is trying his utmost to sound casual, the Kotohana by the woman is glowing red in reaction to his voice, communicating the emotion he is trying to hide. Like emoticons, this product operates "between the lines."

Much research has been done on speech-recognition technology in recent years. The more research we do, the more we find about the unfathomable depth of human speech. It is a difficult field of study, and progress is slow.

A standard method used in speech-recognition technology is for a machine to translate human speech into a line of letters. From there the machine must decipher the context hidden in the letters, form words and finally, in the case of Japanese, match the appropriate *kanji* and *kana* to the words.

The wonderful thing about Kotohana is that it focuses on the emotions that are impossible to translate into words. On top of that, it cleverly communicates results in colors, not words. Even if a machine could correctly read emotions but spelled out its findings in words on a cold display screen, its users might feel alienated (though perhaps not, if the machine used pictograms).

Similar technology is used in a service called Emotion Analysis Talk that deciphers subtle emotions and psychological cues from a person talking on a cellphone. The technology relays what the speaker was feeling two to three minutes after the conversation ends.

The technology behind Emotion Analysis Talk was developed by an Israeli corporation called Nemesysco. It used advanced psychoanalysis technology for security enforcement purposes. The technology was intended for use by the police and military of Israel and Russia as a lie detector in interrogations.

Once a militaristic technology developed for those living in a tense environment travels to the peaceful island country of Japan, it evolves into something frivolous (but in a good way). There is no need to feel shame in this. We can take pride in creating such a product by combining a sophisticated technology with a traditional mentality.

Machines that lead meetings

Engineers in the speech recognition field are trying hard to develop distinctly Japanese products. One is NTT's Meet Ball, which incorporates innovative speech-recognition technology. A spherical object the size of a soccer ball is designed to hang from the ceiling of a meeting room and listen in on the attendees.

Meet Ball does not convert speech into letters but it can recognize who is speaking at any given moment. Suppose an executive is dominating the meeting and upsetting the other attendees. Meet Ball can say, "Excuse me but you're talking too much—you've been speaking 85 percent of the time!"

The machine can say things others in the meeting might not. Meet Ball could conceivably lead the meeting by saying such things as "What does Mr. X think about this matter?" It could even encourage a group whispering to each other to share what they are saying with the room.

The ultimate goal of Meet Ball, its designers say, is to display information relevant to the ongoing discussion at the meeting table. Getting there will be quite an undertaking.

Meet Ball smooths communication while keeping within set boundaries (it doesn't get involved with the content of the meeting). It's a product more for Japan than the U.S., where forceful self-expression, not emotional harmony, is valued in meetings.

Love thy neighbor

What about those machines that, while convenient for the user, are annoying to the rest of us? Engineers must find ways to mitigate the annoyances, and Japanese engineers are doing just that in the field of noise-insulation technology.

Japanese makers are developing noise insulation for home electronics products. Owners of laptops, which aren't generally considered noisy, used to complain that the cooling fan was too loud, motivating manufacturers to install cooling fans similar to ones found in cars.

Hitachi was the first in the world to start installing these in 2002.

Quiet vacuum cleaners and washing machines developed by Japanese makers can be used late at night without bothering the neighbors. The very thought of disturbing someone is enough to unsettle a lot of Japanese.

Did you know that noise insulation for refrigerators and air conditioners is different from those in vacuum cleaners and washing machines? The former is designed for the individual user; the latter, for the user's neighbors. In Japan, making sure your neighbors are not annoyed is a high priority.

One Japanese product in this line that particularly impresses me with its thoughtfulness is the noise-cancellation slippers. They were probably invented because somebody felt guilty about their loud footsteps bothering the neighbors downstairs.

Japanese air conditioners also have clever mechanisms that reduce the noise of the external compressor unit usually set outside on the balcony. They ease the consciences of considerate folks who might otherwise hesitate to turn on the air conditioning at night for fear of waking the neighbors. Daikin Industries developed an air conditioner called Wild Select Multi with a nighttime noise reduction mode that can lower the sound made by an external unit anywhere from three to nine decibels. This kind of sensitivity is a Japanese virtue, though—let's be honest—not all Japanese share it.

These products have been developed for the Japanese market because enough considerate customers value these functions. The Japanese should be proud that such a demand, which indicates a certain refinement, exists.

Noise reduction in earphones

Even the iconic Sony Walkman was born out of a desire to not bother others. Back when Sony co-founder Akio Morita developed the product, he wanted to create a device where you could listen to music in public without imposing that music on passersby. Morita made the Walkman for a Japanese society that tends to follow agreed-upon (if not legally enforced) public behavior. He had no idea that the music device would become a global hit.

In crowded Japan, many of us have come to hate the annoying noise leaking from the earphones of portable music players. Japanese manufacturers have developed new types of earphones to address this problem.

But putting credit where credit is due, it was American audio-equipment manufacturer Bose Corp. that was the first to market noise-cancellation earphones. This is a high-tech product that not only passively prevents outside noise from seeping in, but also detects ambient noise and sends out an inverse (opposite) sound wave to cancel it. This is called active noise cancellation technology.

Still, while researching Japanese manufacturing for this book, I found myself feeling grateful for the cultural foundation underlying our distinctly Japanese style of communication, with its consideration for others.

Japanese culture has long been compared to Western culture and mocked for its immaturity. General Douglas MacArthur notoriously called the Japanese "a nation of twelve-year-olds," and Japanese themselves often say that they have to quickly grow up and become fully mature adults. When it comes to manufacturing, however, the immature Japanese have an outstanding aptitude for making products with potential.

Comic artist and illustrator Masamune Shirow's *Ghost in the Shell* was said to be the basis of the Keanu Reeves hit *The Matrix*. Shirow is a great example of a

Japanese creator. His work is characterized by a deep philosophical world view as well as carefully thought-out and finely detailed machines and tools. In one of his works, *Appleseed*, humankind has grown tired of never-ending war and constructed the utopian city of Olympos, where genetically manipulated humans called biodroids and normal humans live side by side. Biodroids are equal to humans in intelligence but have controlled personalities that prevent them from becoming overly emotional or aggressive. They are in charge of keeping the human population from clashing. Shirow's dream of an artificial product functioning as a buffer and link between people is truly inspiring.

Eliminate Embarrassments: Modesty Pursuits

A naturally modest people

Humans are embarrassed when things they want to hide are revealed. They typically want to hide their feelings, which they hold within the recesses of their hearts or in the gap between their false, self-centered selves and their true selves.

People may also feel embarrassed when for some unexpected reason they stand out from a group to which they belong.

Japanese people in particular have a pronounced tendency to feel embarrassed. Non-Japanese, observing this behavior, may wonder, "Why are they embarrassed by that? Are they children?"

I will try to answer those questions using, as examples, several Japanese products designed to ease embarrassment.

An active ingredient in konnyaku *can stop those embarrassing stomach growls.*

How to get rid of "middle age man's smell"

A special subset of products deals with physiological phenomena embarrassing to the Japanese, but not necessarily toilet-related.

Take the snack Gupita by Narisup Cosmetics. It meets an interesting demand for a product that prevents the noises made by an empty stomach. Loud stomach growls can be embarrassing, especially when dealing with customers or taking part in a meeting. If a supervisor's stomach starts making sounds while he is lecturing his subordinates, his authority is instantly undermined.

With an active ingredient derived from *konnyaku*, a gelatinous substance made from devil's tongue, Gupita expands in the stomach, stopping growls before they can start. *Konnyaku* has almost no calories, so it can also serve as a diet food.

A healthy body produces potentially embarrassing sounds and smells. People tend to worry about smells from flatulence, underarms, feet, scalp and breath. But as soon as you drop your guard, your orifices start emitting odors beyond your control!

Body odor has never been as serious an issue in Japan as it is in the West. For one thing, the Japanese diet has always been centered on vegetables, and it is customary to bathe regularly. For another, close physical contact in the form of shaking hands or kissing is not common in Japan.

Consequently, the market here for body-odor products has never been very significant. Even so, the Japanese have developed this market with a notable attention to detail. One example is products that mask the smell of aging.

Japanese have long commented on the peculiar smell of middle-aged men.

Also, it has been scientifically proven that men in this age group really do give off a certain odor. The source of this middle-aged male smell was discovered in 1999 through collaborative research by Shiseido and Takasago

International. A fatty acid called 9-hexadecenoic acid (palmitoleic acid) is secreted from the epidermal fat of middle-aged men. This substance breaks down into nonenal, which makes the smell associated with aging.

Men secrete almost no 9-hexadecenoic acid until they are in their thirties. But after they turn forty, the amount suddenly increases. This aging smell is more conspicuous in men because women don't have as much epidermal fat.

Shiseido discovered that a combination of antioxidants and anti-bacterial agents eliminated the smell. It marketed a shampoo and cologne containing this compound. Deodorant sales in Japan have traditionally been small, but Shiseido coined a new term, *kareishu,* or "old age smell," for the purpose of cultivating a new market from scratch.

In 2005, Shiseido rival Kanebo Foods, now Kracie Foods, launched Fuwarinka, a product that tackles old age smell from a completely different angle. One or two hours after being consumed, this soft gumlike candy emits a sweet scent through the pores of the body. When humans eat strong-smelling foods, perspiration excreted by the sweat glands carries the smell: When you eat garlic, your sweat smells like garlic. Using this mechanism,

Kanebo's Fuwarinka gum emits a sweet scent through the pores of the body.

Kanebo Foods developed a product with a fragrant substance that exits the body through the sweat glands.

Fuwarinka was targeted at women in their 20s and 30s, but after it went on sale men worried about middle-aged smell bought it too, making the product a surprise best seller.

Then in July 2006, a chewing gum for men, Otoko Kaoru, launched and sold out immediately.

Everyone can understand the idea of chewing gum to prevent bad breath, but the idea for a gum that emits a pleasant scent through the sweat glands was not as obvious. To develop this product, researchers had to thoroughly study the body's metabolism. Their depth of technical knowledge is impressive. I guess you could say they have mastered "the way of gum."

Eliminate Embarrassments: Modesty Pursuits

The "girly" psychology of the Japanese, with its hyper-sensitivity to embarrassment, has changed the parameters of products for middle-aged men, while expanding the market for men's beauty products and cosmetics in Japan.

Hairless men

Of the five senses, the one that gathers the most information is sight. Products that deal with visually embarrassing physiological phenomena can accordingly be very sophisticated.

One such product is *tekari* paper used to absorb skin oil.

Skin excretes more oil, particularly on the forehead and the "T-zone" area around the nose, at night. Both women and men have to watch for oily skin. The Japanese have created a variety of *tekari* products for these needs.

The underarm perspiration pad is another product that addresses visual embarrassment. Commuters on rush-hour trains hanging on straps worry about perspiration stains under their arms, so women's summer clothes often have underarm pads to prevent perspiration stains.

Hair grows in the underarm area, too. The Japanese have relatively little underarm hair, but Japanese women are extremely embarrassed to show even a small amount. When Japanese watch international sporting events like the Olympics, they are always surprised to discover that some foreign female athletes do not shave their armpits. Here it would be unthinkable.

Some young Japanese men have gone one step further, shaving or otherwise removing their body hair. Hair on the chest used to be a symbol of manliness, but now many young Japanese women just consider it weird. Since they prefer smooth skin, body hair has become something even men worry about. Shaving the underarms completely is still considered overdoing it, though. It's more stylish to thin the tangle. Aesthetic salons for men, which permanently remove hair using lasers, are springing up everywhere and prospering.

Men's cosmetics and beauty products and services have been around for centuries. According to manga artist Hinako Sugiura, who researches Edo Period customs, the men of Edo (feudal-era Tokyo) thought it unseemly to let stray pubic hairs stick out of their loincloths. In the public bath, they would even hire someone to pluck the hair from their backsides. In Edo, a city dominated by

Edo Period men were as image-conscious as today's metrosexuals.

the samurai class, women were few, so young men had to work hard to find brides. They would try to make a good impression by smoothing their skin with facial treatments. Just shaving wasn't enough, since the stubble irritated women in face-to-face encounters. Many young men actually plucked their beards.

These men were also anxious about bad breath, so they often brushed their teeth and gargled. They even chewed the buds of plum trees to freshen their mouths before they went out on the town. These sophisticated men of Edo had much in common with the so-called *ikemen* (handsome young guys) of today. I wonder if the older people then disparaged their primping, saying, "Young men today are so wimpy."

Feelings of embarrassment aren't limited to the

body. Take the product known as the silent binder for workers nervous about making a loud *ka-ching* sound when opening a binder during a meeting. Or the silent keyboard, the click-less computer mouse, and the quietly retractable ballpoint pen.

The most unusual examples in this product class are ringtones that duplicate real-life ambient sounds for people who become flustered when their cellphones ring in public. They camouflage the conventional incoming call alert with typical city sounds. To try such an application, simply select Iromelo Mix from a ringtone download site.

Ringtones that mimic the sounds of everyday life— the clearing of a throat, the clicking of a pen, the whoosh of a restroom hand dryer, the hum of a copying machine or the beeping of a fax connection—allow people to use their cellphones in places where a ringing phone might be considered rude.

The concept of "hidden extravagance"

The Edo dandy who plucked his beard or pubic hair in order to attract women or the modern-day *ikemen* who pays for laser depilation can be seen as products of a culture of *notenki* (a hard-to-translate word for a carefree, thoughtless, insolent attitude traditionally considered unmanly) now rampant in Japan's nearly conflict-free society. Similar urban phenomena can be found in other countries, where these men are sometimes referred to as "metrosexuals."

In Japan's peacetime culture, a desire for greater personal attractiveness has become stronger for both sexes. This sort of self-consciousness has a direct relationship to embarrassment: If an *ikemen* can't achieve his desired level of attractiveness, he will be embarrassed to show his face in public.

Let's look at the core psychology of the easily embarrassed Japanese. Children tend to be unsure of themselves in front of adults. They are typically shy. If an adult scolds the child, saying, "Stop squirming and mumbling," the child may become even more introverted and stop talking altogether.

The situation is similar when Westerners speak to Japanese. Westerners present their opinions forcefully based on clear logic, while Japanese tend to be reserved. Under verbal bombardment from Westerners, the Japanese may become bashful and silent.

Because they are inexperienced in social interactions, children and teenagers are often embarrassed by things that adults find trivial. Their egos are immature, so they try to present themselves as better than they really are. Their embarrassment springs from their anxiety about the gap between the image they want to project and their true selves.

When we are young, many things embarrass us, but as we grow old we feel less need to demonstrate our attractive qualities. That's why old people tend to take a defiant attitude, thinking there's nothing to be afraid of anymore. Of course, likable old people exist, such as elderly women who have not lost their youthful charm, but many old folks turn cranky.

For young people who want to win popularity or show their originality, it's normal to spend money on cosmetics, hair care, and clothes to present an attractive face to the world.

The Japanese used to buy expensive clothes—even if it meant spending beyond their means—to look good. But today's Japan is more frugal, and many people are embarrassed to be strutting around in expensive, new clothes.

In vogue these days are clothes that look like they've been worn, with purposely stained, burned, or damaged fabric. This is also an extravagance of sorts—deliberately ruining new clothing for the sake of fashion.

Chrysanthemums *by Ogata Korin.*

Then there are people who don't think it's cool to buy processed used clothing but still want to avoid embarrassment. They may lack confidence in their fashion sense. So-called total coordination consulting services, which help such people select complete wardrobes, have become a viable business in Japan.

An interesting story called "The Extravagance of Korin" illustrates this variety of Japanese vanity. Ogata Korin was an artist who lived in the Edo Period. A man of taste, he later perfected a technique for making beautiful gold lacquerware. Once Korin was invited to a fireworks-viewing party. Though famed for his extravagance, he brought only a simple snack of rice balls wrapped in bamboo bark. At first, the other members of the party laughed at his humble meal, but when they looked at the discarded bamboo wrapping they noticed, to their surprise, that the inside was coated with gold lacquer.

This story illustrates the concept of hidden extravagance, which became absorbed into the general culture over a period of more than two hundred years, during the Edo Period. True extravagance, according to this concept, is refined, but out of sight. Sometimes it takes the form of pricey clothing made to look old and worn.

Conversations between cellphones?

The unassertive and easily embarrassed Japanese often have trouble expressing themselves. But some try to overcome their embarrassment with props, which has given rise to a new type of business for *hitori-sama* (literally, single customers), who are mainly young, unmarried women.

Marketing schemes called "single-person plans" are designed for restaurants, hotels, or package tours. These single-use plans, which offer a touch of extravagance and luxury for single women, have recently been offering more variations. Many are devised to allay embarrassment.

For instance, when eating out, women in general and Japanese women in particular tend to feel embarrassed if they are eating alone.

Two circumstances can typically put these women at ease: (1) All the other customers are also alone; or (2) their aloneness is hidden from the outside world.

Operators of restaurants in office areas once believed they needed to be at street level with picture windows that revealed the interior. However, *hitori-sama* do not want their colleagues to see them through the window eating alone. Consequently, a successful business model has emerged that targets the growing number of single diners by locating restaurants either on the second floor or in the basement of a building, so that no one can spy the diners from the outside.

Products and services that ease embarrassment while creating new types of communication have a lot of potential in Japan.

Passengers in elevators in residential buildings often fail to greet each another and feel uncomfortable for the duration of the ride. If one passenger has a child with her, though, others may try to break the ice by talking to the child. The child thus becomes a prop for communication, if "prop" isn't too blunt a word.

Dogs and cats can serve the same function. If two people both like dogs but don't know each other, they might exchange a few words on the subject of dogs.

A story has recently been going around that carrying a puppy makes it easier for a man to pick up girls. True or not, this belief has caused a boom in the pet rental business in Tokyo's trendy Ebisu and Daikanyama districts.

Regardless of their effectiveness as pick-up tools, props open up opportunities for communication. But other than children or pets, what props can be used to break the ice in everyday situations?

Look to your cellphone, which is evolving into a device with a brain of its own. Cellphone technology is shifting away from mere communication. Say you are in an elevator with another person and the atmosphere is tense. To mask your discomfort, you take out your cellphone and start checking the display or pretend to operate it. These actions send the unspoken message that "I'm busy now so I can't talk to you, but it doesn't mean I'm avoiding you."

If cellphones become smarter, they can be used to bring people closer together.

Imagine this scenario: Two people are sharing an elevator. Their cellphones connect to each other wirelessly and start a conversation: "I met you here at midnight three days ago. Do you also work late?"

This may sound strange, but if a conversation

The virtual pet Tamagotchi has evolved into a communication device.

between cellphones can lead to genuine human communication, what's wrong with that?

Such device-to-device communication is already a reality: Bandai's second-generation Tamagotchi toy Return! Tamagotchi Plus has a built-in feature that wirelessly exchanges "presents" with other Tamagotchi.

This toy will no doubt lead to the production of new types of cellphones and other devices that create new modes of communication by breaking down walls of embarrassment. Many will be for the Japanese market, where there is a strong demand for them.

Ten Geeky-Girly Rules of Japanese Products

Keep It Clean (and Healthy): Avoiding the Icky Sticky

"Adorable" people pursue healthy, long lives

Is the Japanese character, with its mix of shyness and consideration for others, coarse or refined? Those who consider the feelings of others while shyly refraining from emotional outbursts are obviously refined, aren't they?

But there's always a flipside. While the Japanese value harmony with others, there is another national trait that is not so praiseworthy: the selfish pursuit of luxury.

But even when Japanese selfishly pursue luxury, they do it a bit differently from Westerners.

To understand the differences, you need to understand this: Many Japanese think of themselves as *kawaii*, which translates variously as "cute," "adorable," or "precious."

"Self" in this case comprises both the physical and mental.

Japanese feel that their body is *kawaii* or precious to them in two ways: (1) They like their body's external beauty; and (2) they are concerned about the health of their body.

One good example of the former sentiment in action is women's fashion. Japanese women pour tremendous energy into looking their best in public. The amount of money and time they devote to everything from beauty products, makeup, and hair styling to textiles and accessories is head and shoulders above the international norm. And those concerned about the health of their body hope to live a long life without illness.

After World War II Japan built the world's second-largest economy while achieving the world's longest average lifespan. Compared with their ancestors in poorer times or people in developing countries, today's Japanese live in an earthly paradise. Beginning with Chinese Emperor Qin Shi Huang in the 3rd century A.D., the rich and powerful have often sought perpetual youth and longevity. For them, health and long life are an extreme form of narcissistic luxury. Today's Japanese are no different.

Which Japanese products exploit this desire for health and long life? Let's take a look.

Exterminate bugs without touching them

Japanese are the world champion clean freaks. Missionaries who came to Japan from Europe in the 16th Century marveled at the high level of sanitation and hygiene and the love of cleanliness. Our ancestors must have known that cleanliness and good hygiene promote good health and long life. That awareness is still very apparent today.

Some of the products on the market take this penchant for cleanliness to extreme lengths.

In 2003, Chugai Pharmaceutical released a product called Barusan Bubble Hardener. This was a spray for exterminating cockroaches and other insects. Users sprayed liquid at cockroaches with a pressurized can similar to those for ordinary bug spray, but with one big difference.

Barusan was not actually a bug spray but rather a "bug capture spray." With conventional sprays, the user could kill cockroaches simply by spraying them with insecticide, but then had the unpleasant but necessary task of disposing of the corpse. For most people, just looking at a dead cockroach is repulsive. As for touching one, as a Japanese high school girl might say, "Yucky—no way!" Barusan was developed in response to this sensitivity.

Barusan sprayed a bubbly gel that covered the cock-

Chugai Pharmaceutical developed a product that kills cockroaches and encases them in a hard gel so we don't even have to touch their dead bodies.

roach and quickly hardened, thus obviating the need to touch the corpse. This was a new concept for cockroach extermination. Barusan also worked on spiders, stink bugs, ants, pill bugs, woodlice, millipedes, and centipedes.

Cockroaches are universally seen as repulsive. Hardly anyone actually likes them. And while no one enjoys disposing of a dead cockroach, only a rich, peaceful land like Japan would come up with a product that eliminates the need to touch those dead roach carcasses.

Barusan may become useful in other ways in the future. When space travel becomes common, cockroaches may slip into spaceships and breed. If you were to use ordinary bug spray in the weightless environment of a space ship, the poisonous ingredients would form into droplets and float about forever. But with Barusan, you can exterminate noxious insects without worrying about that scenario.

But Barusan may not have to wait until the dawn of the space age to find new uses.

The bottom rails of the aluminum window frames in Japanese homes often contain the dried remains of flies, spiders, and other insects. If the Japanese passion for cleanliness becomes even more extreme, housewives may zap them with Barusan to dispose of them. Also, people who don't want to touch trash or house dust may turn to Barusan. Though these projections may seem funny now, they could well become realities.

Disinfecting the sandbox sand

Some Japanese "clean freak products" are dedicated to the elimination of microbes. The world is already over-flowing with hygiene products and services that come with such catchwords as "disinfect," "sterilize," "anti-bacterial," and "anti-mold." In Japan the number of these products soared in the latter half of the 1980s.

At the same time, some Japanese expressed con-cern that this "anti-bacterial boom" had gone too far. They feared that children surrounded by too many antibacterial products would grow up with weak immune systems.

The Japanese aversion to microbes is deeply rooted. The media regularly works itself into a frenzy over new types of scary bacteria. When MRSA (Methicillin-resistant Staphylococcus aureus), a super-bug that couldn't be killed with ordinary disinfectants, started spreading in hospital wards, it became big news.

The first anti-bacterial boom has passed. But no other people worry as much about mold and bacteria as the Japanese. Our passion for cleanliness is a big part of our national character. To people in other countries, it may seem a kind of obsession.

Services that kill bacteria lurking in park sandboxes have emerged in response to Japanese mysophobia. Various technical approaches have been developed. Instead of heating the sand, workers spread antiseptic on it or replace the sand with anti-bacterial material or mix in artificial sand treated with antibacterial agents.

The demand for these sandbox services has arisen not only from fear of germs in the sand itself but also from the concerns of many mothers about pet droppings, especially cat droppings, which become mixed in the sand and make it unfit to play in.

A few years ago, a rumor went around that eggs laid by roundworms in the bodies of cats could enter children through cat droppings and cause them to lose their sight. Though perhaps just an urban legend, this story gave a boost to sandbox disinfecting services.

But some people counter that untreated sandbox sand can strengthen children's immune systems. While directly learning the difference between clean and dirty (and becoming covered with sand in the process), they build up their immunity to disease through contact with bacteria in the sand.

While understanding parents' concerns, I also believe that sandbox disinfecting services should avoid exaggerating the dangers of untreated sand. Such fear-mongering can be harmful.

But rightly or wrongly, products and services that exploit the Japanese dread of bacteria have an excellent chance of becoming hits.

Plastic is an easy material to make antibacterial: Just add a small amount of antibacterial agents to a mix of coloring pigments and plastic.

Since they're so easy to make, antibacterial plastic products are common in Japan. They are used like ordinary products for the toilet, bath, and kitchen—environments where slime and mold can flourish. Among the more common products are antibacterial plastic cutting boards, triangular sink strainers, bath mats, and shower hooks. Some products, though, are almost too extreme to be believed.

Take the antibacterial rubber band, for example. When you use one to fasten a package of vegetables, you can be sure that the veggies will stay clean. I have to take my hat off to a company that could find a way to improve a product seemingly so simple and fully evolved.

Unicharm has been stressing for years that its tampons are sterilized, a claim that has boosted sales. This example again shows how manufacturers can make mature products stand out from the competition by catering to consumers' antibacterial obsession.

Among other extreme products for daily use are antibacterial bankbooks, bankcards, erasers, and music CDs. Even their designers may not know why these products are necessary.

For the accountant in your life, there are electronic calculators made of antibacterial plastic. Or perhaps they are for "office ladies" (Japanese-English for female office staff), who recoil at the thought of the icky germs on the calculator after the department manager has used it.

When I was still working as an engineer, I once mixed an antibacterial agent into the hard coating used for the surface of a personal digital assistant made for export to the U.S. I remember seeing the sentence "Its antibacterial properties make it clean and safe to use" in the English-language catalog. I wonder what the American businessperson who used the PDA thought of that!

An insatiable passion for cleanliness

Japan also has "super products" for attaining the ultimate level of cleanliness. These are the sorts of products that the less clean among us may find pathetic and obsessive.

For example, there is a product for cleaning fingerprints and skin oil from cellphone display surfaces.

Some people don't even want to touch the cleaning solution when they wipe the icky, sticky display. In response to this demand, Lion released in 2000 the Top Power Tablet, a tablet-type cleaner, while rival Kao followed with Attack Sheet Type, a sheet-type cleaner. Both were for one-time use.

To many Japanese consumers, ordinary cleaning products, which are usually either powders or thick liquids, do not feel good to the touch. Top Power Tablet and Attack Sheet Type supply the need for "no touch" cleaners.

These products have not become big hits, because the average consumer in Japan is not quite that obsessed with cleanliness—but don't count these products out in the future.

Back in the 1980s, during Japan's Bubble economy, the media was full of stories about young women who used chopsticks to put their fathers' underwear in the washing machine, since anything touching Dad's skin was dirty.

Japanese appliance makers, as you might expect, came up with a product to serve this need: a small tub that fitted inside the washing machine so that Dad's and Daughter's laundry could be washed separately. I promise I am not making this up!

While this sort of behavior may seem odd, if Japanese companies just dismiss our human quirks, they'll never launch new businesses. Instead they should accept these oddities as reality and develop products that play to them. It's easy to call this kind of cleanliness a fixation or obsession, but even obsessions can give birth to a new culture.

Heal yourself with electronic gadgets

While Japan has lots of germ-killing products to keep the dirt at bay, it also has a slew of products that make you feel better.

One example is the air cleaner, which is basically a device for filtering lice, mites, grit, dust, bad smells, and other things bad for the body and restoring the air to its original purity. But how about an air cleaner that releases your favorite fragrance?

Natural fragrances have various health-promoting effects, depending on the user's body type and physical condition. Adding them to the air cleaner turns it into a health-care product.

A function that adds positive elements, such as a healthy fragrance, to a product is called "plus sum." Similarly, a function that removes negative elements, such as an ingredient that detoxifies the body, is called "zero sum."

Let's take a look at some of the plus-sum products made in Japan. Through trial and error, Japanese researchers have dreamed up incredible functions and incorporated them into products for mass production.

The "minus ion boom" that began around 2000 has already passed its peak. For a while, though, a wide variety of products were made incorporating this plus-sum function.

Some of these products, such as air cleaners, air conditioners, hair dryers, and clothes dryers, circulate air. Others, such as clothes, pillows, furniture, and accessories, release minus ions mixed with tourmaline into the air.

Minus ions are found in IT products as well. For example, the Prius Deck desktop PC that Hitachi launched in 2002 features a minus-ion generator that uses the corona discharge method. The first PC to have a minus-ion function, the Prius Deck releases minus ions from the back of the main unit, stimulating the user's bodily functions and immune system while relieving stress and exhaustion. Hitachi claims that this PC can help users improve their ion balance and lead a pleasant PC life, to put it in Japanglish.

As the minus-ion boom fades, other health-promoting electrical products are being developed.

For example, a Panasonic air conditioner has the astonishing function, as the company explains, "of enriching oxygen to preserve the ideal oxygen level in the room." Using a special "oxygen enrichment" membrane, the air conditioner produces air with a high 30 percent concentration of oxygen and releases it into the room. A room sensor detects the oxygen level and keeps it at the 20 percent level optimal for humans. Panasonic claims the air conditioner boosts concentration, helps a person's ability to calculate, and reduces the number of mistakes.

In 2005 Daikin Industries launched an air conditioner that releases vitamin C and hyaluronic acid, ingredients used in beauty products said to be gentle on the skin. These ingredients are impregnated in an aeration filter and then gradually released. The company says clinical tests prove that this air conditioner makes the skin moister. In 2006 Sanyo Electric launched Aqua AWD-AQ1, a drum-type washing machine equipped with an "air wash" function that changes the oxygen in the air into ozone, thereby purifying clothes of germs and bad smells.

The machine electrically transforms oxygen into ozone and sprays it on the clothes inside the drum. Ozone has a strong oxidation effect that dissolves or destroys the cell walls of bacteria, ridding the clothes of bad odors and dirt.

Refrigerator makers are also trying hard to get their products to promote health. In 2004, Mitsubishi introduced a refrigerator installed with a special interior LED whose orange light promotes photosynthesis in veg-

The Nano Walk line of pantyhose features nanotechnology that provides skin care and slimming effects.

etables. This in turn boosts the veggies' vitamin C and chlorophyll polyphenol.

The refrigerator used to have the zero-sum task of keeping food from spoiling. But this LED-installed fridge now has plus-sum functions as well.

In 2006, Mitsubishi released a new model featuring a blue light that irradiates food. Adding this light has improved the fridge's plus-sum functions.

Japanese stockings incorporate plus sum ingredients that can beautify a woman's skin and slim her legs. One brand, Nano Walk for Long, Beautiful Legs, features not only rose fragrance, but raspberry ketone, which has a slimming effect, and squalene, collagen, and rose hip oil, which are good for the skin. Other stockings also contain ingredients said to be good for the health, such as amino acids, seaweed, aloe, vitamin C, and catechin from green tea.

These luxurious functions are designed to please consumers with highly refined sensibilities. To put it another way, these picky consumers are constantly pushing product designers to ever greater heights. The strength of Japanese consumer goods comes from the process of constant testing in a tough marketplace. As consumers continue to make their needs known, the path for Japanese manufacturing will be better illuminated.

rule 8

Transform Life into Theater: Everyday Drama

Rewarding the self

Everyone plays the leading role in the unpredictable drama of their lives. When humans start to harbor narcissistic feelings about themselves, they become inhabitants of an internal world that values the self above all. Knowing this, Japanese corporations have developed products catering to this mentality. Here is another market with great potential.

"Reward yourself, you've earned it!" is the kind of phrase one often reads in women's magazines these days. Often the reference is to a sumptuous meal or a getaway trip to a hot springs. Everyday life becomes theater in this urge to reward the self. The extreme form of this obsession with the self is full-blown narcissism.

Life-as-theater also takes more public forms, as exemplified by *cosplay* ("costume play" or dressing up to resemble favorite anime and manga characters) and maid cafes (coffee shops whose female wait staff dress in frilly "maid" costumes).

The wait staff at a maid cafe play the role of maids serving the whims of their mostly male masters. They play supporting roles in the men's narcissistic dramas.

Until recently, maid cafe wait staff were pejoratively labeled "Akiba-kei" (Akihabara types) and their customers, "*otaku*" (geeks), implying that they belong to a strange subculture centering on the Tokyo electronics district of Akihabara.

Today, *otaku* culture is no longer only for fringe types and social outcasts. Instead, it is integrally related to Japan's "cool" popular culture as it is perceived around the world.

Some dismiss *otaku* as losers who have lost touch with reality because they read too many manga and watch too many anime films and TV dramas. Such critics forget that they too are heavily influenced by fictional worlds and also think of themselves as lead actors in the theater of life, like the patrons of maid cafes.

Giant screens permeate the city

Imagine you are watching a soccer game in a large stadium. The crowd goes wild as it witnesses an amazing play. But what if the crowd were denied the pleasure of re-viewing the play on the stadium's big screen in slow motion? Fans would be frustrated to no end.

The first giant screen in Japan was the Aurora Vision, developed by Mitsubishi and installed in Tokyo's Korakuen Stadium. It gave fans close views of the live action as well as instant replays. Today, sports arenas without giant screens are unthinkable. Even horse racing tracks, a favorite haunt of elderly men, now come equipped with giant screens. In the future, events now open only to the media or select members of the public, such as sumo training or Self-Defense Forces exercises, may also appear on giant screens as a matter of course. This is another example of how reality becomes drama in Japanese life.

Giant screens are no longer limited to stadiums these days. They can now be seen on the streets, blaring advertising to the passersby. Go to Tokyo's Shinjuku Station, the busiest station in Japan, or the famous "scramble intersection" near Shibuya Station, and you will be amazed by the number of screens

Giant screens at Shibuya Crossing in Tokyo.

on the sides of buildings, blending perfectly into the urban landscape.

Tokyo has long had a reputation as a city dominated by neon signs. Beautiful neon signs can also be found in New York and Hong Kong, but in Tokyo they flash rhythmically and move dynamically in sync, adding a distinctive touch to the nighttime scene.

Electronic paper—light, paperlike sheets that can be folded and function as displays—may contribute to this scene in the future. Advertisers may plaster these sheets on telephone poles and walls, with each displaying a moving image—adding more drama to the cityscape.

Even today, advertising plays a key role in dramatizing reality around the world. Since the end of the Cold War, the cityscapes of Moscow and Prague have been transformed beyond recognition due to the growing presence of brightly colored advertising billboards.

In the future display screens, big and small, will be familiar sights in every crevice of every major city. Just as it's hard to imagine New York or Ginza today without billboards, a time will come when pedestrians will feel strange if they aren't constantly surrounded by advertising videos.

Mannequins with anime faces

Fans of *moe* culture are at times stereotyped as loner *otaku* who prefer the two-dimensional women found in anime and manga to their real-life counterparts. The older generation in Japan tends to label these fans as detached from reality and a race apart from the normal majority. But given cultural trends in Japan today, such sentiments are out of touch.

Moe is defined as the fetishized adoration of manga- or anime-like girls. But it is quietly insinuating itself into the wider culture in subtle ways. One way it does this is through mannequins.

In the children's section of Japanese department stores, mannequins look more and more like anime characters these days. Newer models are adorned with the giant star-filled eyes found in the characters in girls' manga and originally depicted in Osamu Tezuka's classic manga *Princess Knight*.

Mannequins dressed in school clothes for elementary schoolchildren are starting to resemble the *bishojo* ("pretty girl") figurines found in Akihabara. Dress up these mannequins in short skirts, pose them provocatively, and place them in the *moe* hot spots of Akihabara and Ikebukuro, and passersby will immediately label them as figurines for perverted *otaku*. But if these same anime-faced mannequins are outfitted in school uniforms and posed innocently holding hands in a department store display, they look like something out of a fairy tale, not to mention being PTA friendly.

Japanese children are very used to seeing children portrayed on TV by cartoon characters. By the time they grow up, only fifteen years or so from now, sensibilities seen as typically *otaku* may become the norm. I believe that the *moe* culture of *otaku*, with its theatrical, fictional world view, is a harbinger of the future.

In addition, anime-faced mannequins hold the key to solving a problem that has long plagued the development of humanoid robots.

When humanoid robots are made too realistic, they create unease in the viewer.

Scientists and engineers have tried and tried to make robots closely resemble humans. It's a tough problem, and no one has figured out a perfect solution. When a robot's face is designed to resemble a human face, initial familiarity quickly turns to unease if the robot is made to look too realistic. Robotics experts call this the "uncanny valley" effect.

This effect has served as the crux of many science-fiction stories, in which robots with human features rarely live peacefully with their makers.

On the other hand, if technology advances to the point of creating truly realistic humanoid faces, we would face the dilemma of not being able to tell them apart from real people.

Because of such issues, many humanoid robots are

made to resemble humans only slightly. Take Asimo, whose face resembles a closed astronaut's helmet.

In the world of fiction, the best-loved robots are not realistically human, such as C3PO from *Star Wars* with his brilliant metallic face. Though bipedal and of human height, he has a uniquely robotic charm.

Meanwhile, anime faces, which used to be found solely in two-dimensional mediums such as paper and TV monitors, are appearing on three-dimensional figurines. The problem of creating a humanoid face that real humans can comfortably live with might be easily solved by using anime models. Mannequins modeled after anime characters are one example of this solution in action.

In Japan, mannequins were long made to resemble Westerners in hair color, eye color, and physical appearance. One reason was the strong Japanese inferiority complex toward Westerners that developed after Japan's defeat in World War II. Only in recent years have anime-faced mannequins, with their non-specific nationality, started appearing. These mannequins signal the possibility of Japan becoming a cultural epicenter in the future.

Gold-medal-winning marathoner Naoko Takahashi listened to the song "Love 2000" by hitomi to heighten her concentration.

The soundtrack to your life

When we star in our very own drama and turn our lives into a piece of theater, we add color to an otherwise monotonous, monochrome existence. Or so the thinking goes.

Just about everybody in Japan knows that marathon runner Naoko Takahashi, nicknamed Q-chan, listened to the song "Love 2000" by hitomi to heighten her concentration during the 2000 Sydney Olympics. This was her theme song for the role she was about to play: Olympic champion. Such music signals that the drama of your life, starring you, can commence. It's similar to the sound of a clapper board on a film set. And in Takahashi's case, it worked very well: She won the gold.

Sony helped unleash this phenomenon of narcissistic theme music—the soundtrack to our lives—piping through our portable music players. I admit I've played my own soundtrack from time to time. I like to play a certain type of music before a special event or a holiday to link the music to the event. Then, when I hear the music elsewhere, it triggers good memories. This life-enriching habit combines two benefits—mental prepping, pleasant recollections—into one action.

Just think of all the possibilities for personal theme music.

Say you are rushing up a flight of stairs to catch a train so you can make an important business appointment on time. Imagine if your phone could calculate your pace and current position to determine that you are moving faster than usual up the stairs. The phone could then choose a theme song for you, say the theme from *Taiyo ni Hoero!* (Shout at the Sun!, a popular Japanese police drama that aired from 1972 to 1986). Hearing it, you could feel the exciting rush of a detective chasing down a suspect. For younger users who don't know *Taiyo ni Hoero!*, the machine could play the theme from *Odoru Daisosasen* (Bayside Shakedown, a hit police TV drama and movie series that started in 1997), *Mission:Impossible,* or another action series. Any of these songs will serve to make you feel like a protagonist in a drama.

Or consider this common scenario: You are on the phone with your friend and happen to say, "It's really hot today." Using its vocal recognition technology, your phone can add cicada noises to your comment. (To Japanese, the sound of the cicada signals the heat of summer.)

Speech-recognition technology, as mentioned previously, is not yet at the level where a cellphone can understand its user's conversation. It could, however, recognize a few thousand or so phrases relating to specific topics like the weather. In addition, it could use its heat sensors to measure the temperature and search the Internet for weather updates to gauge how much discomfort its owner is feeling at a given moment. Such functions will be technologically feasible in the not so distant future. They may feel strange at first, but we will soon start taking these personalized sound effects for granted until a life without them will feel quaint and sepia-tinted.

Subtitled scenery!

Japanese anime and TV dramas incorporate sound effects never heard in real life to enhance viewer familiarity with the show's world. For example, whenever Tara-chan, a three-year-old boy in the long-running family sitcom anime *Sazae-san*, walks, we hear a tinkling sound.

This development could lead to text entering more into our everyday lives. One example is Head Mount Display (HMD)—a computer worn like a pair of glasses. It is the equivalent of glasses lenses that display pictures and subtitles. Professor Masahiko Tsukamoto of Kyoto University, a proponent of the device, has suggested many uses for it.

With the HMD, it becomes possible to project subtitles on the real world. It can detect the mood and feelings of its user and display subtitles based on its findings. Imagine this scenario: A frustrated boss wants to shout at his subordinate when a peaceful-looking penguin pops up on his HMD display screen and tells him to chill.

Another interesting product appealing to self-dramatizers is a music player made to be played in a bathroom developed by Twin Bird Kogyo, an electric home-appliance manufacturer headquartered in Tsubame, Niigata Prefecture. The player features back speakers that bounce sound off the walls, giving listeners an experience of total sound immersion.

When designing sound equipment for the bath,

such basics as waterproofing, anti-slip surfaces, and heat resistance need to be considered. Twin Bird Kogyo has gone one step further by giving bathers the unique enjoyment of echoes bouncing off the walls.

Let's take this idea one step further and imagine a device that can add an echo to sounds produced in the bath. In Japan, bathing still has strong associations with old-fashioned public bath houses. When Japanese set down their wash basin in the bath, they expect to hear the unique echoing "plonk" on the tile floor of the bath house. But in the typical unit bath found in many Japanese homes today, they will probably hear a duller, less nostalgic sound when they put down their basin. But what if they could hear that old-fashioned "plonk" sound reverberate through their home bath or have their humming echo as it would in a bath house in Tokyo's *shitamachi* (old downtown) district?

This is another example of technology that could make our everyday lives more theatrical. It works on the same principle, outlined earlier, of adding sound effects to movies and TV dramas to make them feel more "real" than the real world.

Earlier, I explained a noise-reduction technology that allows engine sounds to be heard at a comfortable level. By developing this technology further, a device in your car could automatically switch the engine noise to that of a Harley Davidson bike when you play "Born to Be Wild" from the *Easy Rider* soundtrack. When you play the *Top Gun* soundtrack, it could change the sound to that of a jet engine. You could even try playing the *Star Wars* theme. The possibilities are virtually limitless. You could transform yourself into the star of your favorite movie whenever you drive.

Dinner with a virtual partner

Next time you sit at home alone, having dinner in front of the TV, imagine having the Come Eat With Me DVD in which a dinner companion appears onscreen to keep you company while you eat. You can have an exotic curry dinner with an Indian or watch a two-year-old try his hardest to eat orange jelly.

The DVD features eighteen dinner partners that vary in age, profession, and other variables such as whether they are left- or right-handed. The menu is also diverse, featuring hamburgers, cup noodles, delivery pizza, and rice balls from the convenience store. You can also choose how much time it takes to consume the food.

For lonely souls, this DVD can be comforting, but it can also cause them to realize their dire situation. More conservative types may slam it as being in poor taste. Some may even become angry at how low we Japanese have sunk as a people.

For me, however, reality-augmenting technology that dramatizes everyday life has great possibilities for the future. I can't think of a country other than Japan currently developing these kinds of products.

Be Good to the Planet: Eco-Friendly/Eco-Product

A leader in major environmental technology

When I work as a technology management consultant, I often come across statistics and materials related to Japan's global competitiveness in industrial technology.

The term "technology" covers many things, from the development of nuclear energy to the commercialization of genes, and many institutions compare the quality of technologies from different countries. As a simple example, let's compare the number of patents registered each year worldwide in different fields.

The Japanese government has devised something called a Second-Phase Scientific Technology Basic Plan 1 for developing strategic technologies in four fields: life sciences, data telecommunications, environmental science, and nanotechnology and materials. The government has decided to invest significant resources in these fields, as well as in energy, manufacturing technology, infrastructure, and frontier technology. It is promoting what it describes as "effective and efficient research and development" with the aim of "obtaining patents for research results" that ensure ownership of intellectual rights in strategic technologies.

In Japan, the U.S., and Europe, about 150,000 patents were registered in the aforementioned eight fields in 2005. Japanese patents numbered 38,000, or about 26 percent of the total. That's about the same percentage as Europe and about half of the U.S. Of course, patents in these fields have also been registered in the rest of Asia and Latin America, but Japan does about one-fourth of the research and development from the three economic leaders, which determine the direction of advanced technology. Japan's economy is about the same size as Europe's, and its technological level is similar in almost all industrial categories.

Now let's compare Japanese patents to those from the U.S. and Europe in each of the designated fields.

Japan received 16 percent of the global patents in life sciences, 25 percent in data telecommunications, 58 percent in environmental technology, 28 percent in nanotechnology and materials, 30 percent in energy,

33 percent in manufacturing technology, 56 percent in infrastructure, and 20 percent in frontier technologies. The most notable statistic is the 58 percent share for environmental technology patents. This clearly indicates not only that Japanese companies are interested in eco-businesses, but that the Japanese people are seriously concerned with global environmental issues, which are projected to become even more critical in the future.

The Energy Conservation Center, a foundation that studies energy efficiency worldwide, has devised an index for energy consumption that compares oil consumption to GDP. Japan's index is 92, Germany's 130, Italy's 140, France's 147, England's 176, America's 253, and China's 827 (unit: equivalent tons of oil used per US$1 million in GDP). This shows that Japan is more energy efficient than the rest of the industrialized world. In other words, Japan creates the most value from the same amount of oil.

The "eco car" as intellectual status symbol

Japanese companies are developing environmentally friendly technologies that use smaller amounts of energy. This effort has led to success in the world market. In the

The Prius, released in 1997 by Toyota, sparked huge demand for hybrids.

automotive industry, it has served as a big engine for recovery.

Of the classes of cars considered environmentally friendly, hybrids are the best known. They're sophisticated vehicles with complex systems, and Toyota and Honda are the only companies that mass-produce them.

The first hybrid car, the Prius, came out in 1997. It took a decade before U.S. automakers began producing hybrids, although on a much smaller scale than Toyota or Honda.

A hybrid has two drive mechanisms: an internal combustion engine that uses gasoline and an electric motor. This complicated power transmission technology cannot be adopted overnight. The hybrid's electronic controls are also highly advanced. Electronics account for as much as

50 percent of a hybrid's total cost, leading it to be called a "computer on wheels."

Hybrid cars have had an enormous impact on the Japanese and American automobile markets. American movie stars and opinion makers purchase hybrids as status symbols, but not of the wealth-flaunting sort. Instead they are symbols of the owners' concern for the environment. Driving a Prius gives them the standing to promote their environmental causes that conventional cars cannot.

The Prius also has an image of being driven by people of high intellectual achievement and deep social concerns. It has thus become an intellectual status symbol: Smart people drive a Prius.

A similar phenomenon occurred in the U.S. in the 1970s, when Honda's CVCC engine filled a role like that of the Prius.

Air pollution standards were revised through the Muskie Act, which mandated that harmful auto emissions be reduced by one-tenth in five years. Honda was the only company in the world that complied with this mandate thanks to its development of the CVCC engine.

The Civic CVCC was popular in both Japan and the U.S. during the first Oil Shock (1973–74), which occurred shortly after the CVCC debuted. Meanwhile, Honda's other four-wheel vehicles recorded markedly improved sales.

The Prius and the Civic CVCC symbolize the rise of Japan's automobile industry. At their launch, both boasted advanced technology whose environmental performance outstripped the competition.

A culture of hospitality and conservation

Environmental technology can be divided into two broad categories: cleaning technologies such as those that eliminate harmful chemical substances or the dangerous effects of electromagnetic waves; and energy-saving technologies that conserve finite energy resources according to the three R's—reduce, reuse, and recycle.

Japanese manufacturers are world leaders in technologies that reduce the burden on the environment. The Prius and the Civic CVCC stand out in terms of not only clean energy, but also conservation because of their superb gas mileage.

JFE Engineering's reduced-emission, high-temperature refuse incineration system and Asahi Glass's fluorocarbon-resolution liquid conversion system are also examples of Japanese clean technologies.

Other conservation technologies from Japan include Sharp's flexible solar battery; Matsushita Electric's home-use fuel battery cogeneration system; YKK AP's wall-cooling technology using optical catalysts; Isuzu Motors' hybrid car systems using electrical two-layer capacitors; and Daikin Industries' low-energy hydraulic pumps.

Japanese recycling technologies include cement made from the ash of refuse incinerators, developed by Taiheiyo Cement, and an advanced super-conductor electricity system that stores energy using electromagnetic balance coils, designed by Chubu Electric Power Co.

The Japanese tend to adapt to their social surroundings by trying to contribute to overall harmony. Thinking only of your own needs is considered low-class self-indulgence. Perhaps the true self-indulgence is consideration for others, because of the feelings of satisfaction and happiness it engenders. The Japanese are respected worldwide for this quality—and they can tap into it to create new products.

This kind of consideration can expand beyond one's immediate circle and community to Japan, Asia, the world and, ultimately, all living things. It underlies Japan's serious concerns for the environment.

Similarly, the Japanese concern for cleanliness,

which some call an unhealthy obsession, has contributed to the development of clean technologies. As I mentioned earlier, this concern is hardly recent. Western missionaries and merchants visiting Japan in the 16th and 17th centuries expressed amazement in their dispatches home at the cleanliness of the streets, the love of hot baths, and the attention paid to sanitation. Impressed by the Japanese attention to cleanliness, Westerners took it as evidence that Japan was an advanced civilization.

Of course, a desire for cleanliness doesn't mean everything is environmentally perfect. Japan has had to address a number of serious industrial pollution problems, from the cadmium-caused itai-itai disease, which first appeared in 1912; to the mercury-caused Minamata disease, whose first victims were diagnosed in the 1950s; to the nuclear meltdown in Fukushima in 2011. Nevertheless, Japan today has a cleaner environmental record than many countries.

Priuses of the future

The Japanese penchant for cleanliness and their desire to work in harmony with others are positive traits when it comes to environmental issues. But what about the nation's tendency to anthropomorphize objects? Doesn't that seem almost anti-environmental?

Businesses have a hard time getting the Japanese to recycle their cellphones. They get so emotionally attached to them that they are reluctant to throw them away.

Foreigners living in Japan often comment on the clutter in Japanese homes. And while sales of older houses and used clothing have been increasing in recent years, recycling is still not widespread here. One reason, I believe, is the Japanese tendency to form emotional attachments to objects.

Many years ago, when Japanese designers were revamping the flush toilet for the local market, they added the "small-flush" function. This design innovation took hold as more and more Japanese installed flush toilets in their homes. An unintended benefit was the conservation of an enormous amount of water.

This happened before global environmental issues were on everyone's mind, but one result of the small flush function was to heighten environmental awareness among ordinary Japanese. That heightened awareness has since become a major characteristic of Japanese culture.

The small-flush idea became the foundation for a

number of environmental technologies that have since been turned into products. The exterior casing of a cellphone launched by NEC in 2006 was made of polylactate, a biodegradable plastic. Conventional petroleum-based plastics are not biodegradable, but polylactic plastics are made from corn. Polylactic acid is combined with crushed wood chips from Africa called kenaf to improve durability. It is an advanced compound.

Polylactic acid was originally used for household garbage bags, commercial food trays, and packing materials, but NEC adapted it to an expensive, high-tech product: the cellphone. It became a vanguard technology that is now used worldwide.

Some may think such a technology is an extravagance, but these products were created for users who want to be seen as eco-conscious. They will buy a biodegradable cellphone simply because it is environmentally friendly.

Japan is a country with a high standard of environmental awareness, so new eco-products with the same market impact as the Prius and the Civic CVCC will no doubt emerge here.

rule 10

Go Small, and Smaller: Downsizing

From revolutionary to the norm

Products and services face the inescapable fate of a limited life cycle. Products with new functions move from a germination period, when they are being invented; to a growth period, when they acquire more functions; to a mature period, when they add new functions that have little to do with the original ones and start competing on cost. Product functions that are initially regarded as earth-shaking gradually become nothing special as users become accustomed to them. In the end, they suffer the fate of becoming "built in," meaning "installed" or "standard." In other words, even revolutionary features become smaller and less prominent with time until they finally become completely absorbed into the product or environment.

The history of the home television receiver illustrates this product life cycle.

In the 1960s, ordinary Japanese households were adding TVs en masse. Together with the refrigerator and washing machine, the television was called one of the "three treasures"—it was a symbol of modern life. The living room TV, designed to attract the eyes of visitors, had a solid presence and dark imitation teak finish, mimicking a heavy, dignified piece of furniture.

After TVs became common, their prices fell drastically, and in no time it became nothing special to have a set in every room. Televisions were no longer a rarity, and consumers came to prefer ones with space-saving features.

TV makers passed through a challenging period from the 1980s, when they labored tirelessly to make a thinner picture tube, to the 1990s, when they started to mass-produce liquid-crystal receivers that led the way to today's flat-panel displays.

As TVs become ever lighter, they can easily be hung on walls, and slowly blend into them. Similar to flat-panel TVs, front projectors with liquid-crystal panels that receive images from distant sites such as movie theaters are gradually being downsized. In any event, TVs are becoming extremely thin. In the end, they will be standard parts of the wall or ceiling.

Air conditioners have also been getting thinner, with the thinnest current model being 14cm from front to back. In the end, the air conditioner will become built into the wall with a thickness of zero.

Vacuum cleaners are similarly undergoing this process. Japanese vacuum cleaners are among the best in the world in terms of having both strong vacuuming power and a small size, but in this case there is a limit to how small the stand-alone product can become.

A "built in" vacuum cleaner is already being developed: The body of the cleaner is installed under the floor of the house. Hoses are run under the walls and floors, and each room is supplied with a suction hole, somewhat like a wall plug. When you want to sweep, you simply insert a handheld hose into the suction hole built into the wall or floor of every room.

The "built in" concept is not limited to objects possessing size and shape. It also exists in the realm of color.

Consider the changing colors of public phones in Japan. After red was formally designated as the standard color for public phones in 1953, red phones stood out in darkened Japanese streets for nearly twenty years. In 1972, just before the Oil Shock, the yellow phone made

A typical payphone in Japan.

its debut. In 1982, a green phone that accepted magnetic cards appeared.

In 1992, the proliferation of pirated magnetic cards spurred the debut of a pirate-proof IC-card phone. Painted a subdued shade of gray, it is still in use.

These transitions are like watching a traffic light change. From red, the color that stands out most, phones became yellow and then green. They finally turned gray, like the traffic signal poles that blend into the urban landscape.

The history of the Japanese mailbox is similar. Since 1945, nearly all mailboxes have been red, but "camouflage" mailboxes painted gray and navy blue have recently started to appear in Ginza and Yokohama to better fit in with the surrounding scenery. The public

Ten Geeky-Girly Rules of Japanese Products

phone and mailbox, which are being pushed aside by the Internet and the cellphone, are being built into the landscape, meaning that their existence is no longer being emphasized.

Wearable products

The vacuum cleaner, the TV, and the phone booth are quickly being built into our landscapes and interiors, but what about technology we wear and carry?

Perhaps the longest-running high-tech device that we wear is the wristwatch. We wear it without a second thought, but the wristwatch had to fight a long battle for acceptability.

The history of the mechanical clock goes back to around 1300. Clocks in that period were the ancestors of today's grandfather clocks—big stand-alone objects. Three hundred years later, around 1600, a downsizing technology had developed in response to the need for a portable clock. The resulting product was called a "pocket watch." (If it had been developed today, it might have been called a "mobile clock.") By the time the pocket watch appeared, precision technology for reducing size and weight had nearly advanced to the level of the wristwatch. But another three hundred years would pass until the first wristwatch appeared—a model by Cartier for men that debuted in 1904.

For people to start wearing such an artificial object on their skin, the barrier to this particular man/machine interface had to be overcome. That this did not happen until the 20th century shows how high the barrier was. In other words, it took three hundred years for the watch to move to the wrist after first earning the right to be in a suit pocket.

Eyeglasses came into use even before the wristwatch. It took about four hundred years—from the development in the 13th century of glasses that were held in one hand, like opera glasses—until glasses could be worn on the ears. This illustrates how difficult it is for manmade objects to achieve the right to become wearable.

The Japanese knack for wearable tech

Technology begins to become wearable or built in when consumers want it to become invisible. They want the technology—be it a wristwatch or a vacuum cleaner—to perform its mundane function and nothing more. This trend toward invisibility is the opposite of the usual trend in Japan, where products are enhanced, tweaked, and anthropomorphized so that consumers can form emotional attachments to them.

One good example of this contrast is the different ways Japanese and American treat their cars.

It's uncommon now, but you used to see cars in Japan with deep pile carpets on the floor because no one wore shoes inside. In the parking lots of provincial supermarkets, you still see sandals that have been taken off and left behind by their wearers. This indicates that Japanese are reluctant to soil the interior of their treasured car with their shoes.

On the other hand, most Americans see the car solely as a tool for transportation. They not only ride in the car with their shoes on, but wash it as little as necessary. The Japanese Sunday ritual of Dad diligently polishing his beloved car with carnauba wax in front of the house barely exists in the U.S.

Japanese and American sensibilities are drawing closer together, but a gap still remains between the American way of regarding the car as a faster pair of legs to get around town and the Japanese concept of it as a precious treasure and family member. When a product becomes mature, users can take what the Japanese call

Sony's Walkman helped pave the way for today's mobile communications culture.

a "dry" or "wet" approach to it. The American attitude toward the car is an example of the dry approach; the Japanese penchant for forming an emotional connection with the car is a wet approach.

If manufacturers take the dry approach, they often end up developing ways to make a mature technology built in or wearable. The existence of the product itself is de-emphasized. Typically, a smaller, lighter product is formed. This part of the dry approach plays to Japan's strengths in downsizing.

Sony's transistor radio and Walkman laid the foundations for today's mobile communications culture. Sony's "passport-size" video camera allows people to record their lives anytime and anywhere. This camera led the way for the current age of ubiquitous computers.

Ten Geeky-Girly Rules of Japanese Products

Nikon and Olympus cameras, Seiko and Citizen watches, Casio calculators, electronic dictionaries, and digital watches, as well as other small, light, high-tech portable products from Japan, have conquered the world market.

The process of downsizing and mass-producing mechatronic products (products developed with technology integrating precision machinery and electronics) is basically compatible with the Japanese temperament. In answering the questions "How should we make products?" and "What products should we make," Japanese manufacturers can blaze new trails by following the path of "small is better."

Origami technology in manmade satellites

One basic approach to downsizing comes from Japan's ancient craft of folding, origami. Many folding products are in common daily use, including the folding baby carriage (Maclaren of the UK, 1967) and folding umbrella (Knirps of Germany, 1928).

In Japan, "the country of rain," the average per capita annual consumption of umbrellas is the highest in the world. I'd like to introduce a very Japanese product that uses umbrella-folding technology.

If you strip the cloth from an umbrella, leaving only the frame, and turn the frame upside down, it becomes a hanger for drying clothes. No one knows who came up with the idea for this hanger, but the inventor of a mini-hanger for travel use—a downsized version of the original hanger—is believed to be of Japanese origin. When Japanese travel, they hook this mini-hanger on the rail of the veranda and use it to dry socks and underwear.

Of course, compared with electrical goods, this is a simple product. But whoever had the idea of turning an umbrella frame upside down and using it as a tool was both subtle and bold.

Chairs, desks, ironing boards, and tents also fold. Folding is useful for tools that are sometimes used, but are ordinarily stored away. Compared with other advanced countries, Japan has the least living space and has accordingly most actively adopted folding technology.

At bedtime, Japanese fold the low table called *chabudai* and set it against the wall. They then take out the futon, which has been folded and stored in the closet. This custom of folding and storing bedding is uncommon elsewhere in the world.

Japan's culture of using wood and paper in daily

A woman wearing a kimono walks through Kyoto's Gion district.

life also has deep roots. The Japanese have developed various objects from these materials that fold compactly, as symbolized by the craft of origami. Among examples are the Kurashiki wrapping cloth, kimono, paper fan, and paper lantern.

Origami techniques are being used in advanced technological fields. The best known of these technologies is the Miura map fold, which is a special folding technique used for solar panel arrays. This space technology was developed by Koryo Miura, an emeritus professor at the University of Tokyo, for the Japan Aerospace Exploration Agency, which is known for the H-II rocket.

It's basically a method for folding a sheet into a small, compact shape. Take a rectangle of paper and fold it one-fourth its length using a mountain-valley-mountain fold. Then fold and refold this strip accordion-style, at a slight angle, and unfold it. After making some adjustments to the mountains and valleys, recollapse it so it folds up neatly. At this point, you should be able to open and close the paper easily by tugging on two opposite corners.

This method in not only easy to use, but keeps the mountains and valleys from reversing once the paper is folded and makes the paper itself hard to tear.

Though the Miura fold is now widely used for portable travel maps, it was originally a technology developed to make solar panels on space satellites more compact when folded. This technology, which enables the storage and easy operation of large solar panels in the limited space of a satellite, has its roots in the culture of origami.

The Japanese use origami methods to make even ordinary, non-advanced products. One example is the folding technology for flexible circuit boards and wiring. Electronic circuit boards used to be made of a hard board on which parts were soldered, but the latest circuit boards and circuits are made of soft materials. No parts are soldered on; instead there is just soft wiring designed to bend.

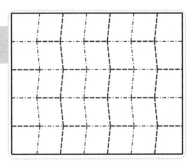

The body of a video camera has various switches and sensors, as well as display lights and ports that hook up to external devices. Inside, the body wiring is needed to connect all the parts. Since the video camera has been downsized to the limit, the wiring has to thread tiny openings among all the crammed-in parts. That's when paper-folding technology shows its mettle in the perfect bends of the camera's flexible circuit board. In addition to flexible circuit boards bent at various angles inside, the video camera itself has a folding mechanism: The screen display that folds open. Japanese maker NEC was the first to incorporate this mechanism for opening and closing the screen display into a cellphone. This style of phone has since become the mainstream but was a fairly radical idea at the time.

Go Small, and Smaller: Downsizing

Origami robots

The Super Alloy Robot was once all the rage with Japanese boys. Its predecessor was Mazinger Z, a robot hero in an early 1970s anime. Bandai revived it in an exquisitely crafted metal model. Later the technology for manufacturing toy robots advanced further, producing the popular transformer robots.

The transformer usually changes into a plane or armored vehicle. But when a tough opponent appears, it combines with its companions to become a giant robot in human form.

The Super Alloy Robot was this sort of model. Of course, animators didn't have to think about designing a detailed structure for it, but the designers of the toy did—a painstaking task.

Today the designing of toy robots starts at the anime stage, incorporating transformation and combination.

The toy that finally arrives on the store shelves transforms using a bizarrely complex folding mechanism. Strangely connected to origami, this object inspires simple astonishment as one wonders how the transformations are accomplished.

Finally, let's take a look at the cardboard box in which the video camera or transformer robot are packed.

Origami cranes.

Styrofoam has fallen out of favor due to environmental considerations. Cushioning packaging, made from cardboard folded into a three-dimensional shape, is now often used instead.

To improve production efficiency, this cushioning packaging is ingeniously made by folding just one sheet of cardboard. It's a fairly tricky process. With a video camera, for example, the attachments and manual must be packed alongside the camera itself. To do all this, an extremely complex design is required, turning the package into an example of origami craftwork.

Designers could not achieve the excellent shock absorption of Styrofoam just by folding cardboard, so they had a tough time at first. Through the process of trial and error, they developed a technology that could pass the drop test, which is when package contents are dropped from a certain height to see if they can withstand the fall.

Japan's traditional origami culture, in the form of a folding technology for packing, is used for everything from the interior wiring of high-tech devices to a new packaging technology with open-and-shut parts.

part 3

Embracing Our Inner Geeky-Girly

chapter 3

What It Means to Be "Girly" and "Childlike"

About those ten rules

The ten rules of Japanese product creation outlined in Part 2 of this book can be divided into two groups: rules that come from the Japanese ability to relate intimately to tools and machines, and those that come from a sophisticated culture that allows self-indulgence in its purest form. Also, rules 1 (Make the object [almost] human), 2 (Pursue personalization), 3 (Create compulsion), and 4 (Leave *something* out) all come from a specifically Japanese understanding of the world.

The Japanese are supremely talented at:

Anthropomorphization: They can become so emotionally attached to their machines and tools that they hold funeral rites for needles.

Personal customization: They not only prefer personal tableware (a personal pair of chopsticks, rice bowl and so on) but also feel a close partnership with tools and other objects. This is the sensibility that affectionately gives bubble wrap the cute product name Puchi Puchi. It also prefers to stop short of perfection, as seen in the Japanese engineers who, instead of making a car interior completely silent, transformed engine noise into personalized sound effects.

These traits all come from the Japanese mindset of developing close, personal relationships with tools. This is akin to a child's fanciful worldview of seeing everything, even objects, as potentially human.

The remaining six rules—5 (Connect us with kindness), 6 (Eliminate embarrassments), 7 (Keep it clean [and healthy]), 8 (Transform life into theater) , 9 (Be good to the planet), and 10 (Go small, and smaller)— derive from the sophistication and ripeness (or, if you prefer, the decadence) of Japanese culture.

Japanese are so obsessed with cleanliness that they have anti-bacterial checkbooks. On the other hand they are decadent enough to make mannequins with anime faces. They also make snacks that stop stomach noises and develop a device that allows one driver to thank another for an act of kindness by waving a mechanical tail. These traits are a world away from selfish consumerism. Instead they evidence a feminine delicacy and sensitive, caring heart.

At the root of Japanese manufacturing lies a feminine delicacy and shyness as well as a childlike curiosity and fantasy-filled worldview. In other words, a "girly" (youthful, feminine) mentality is behind the unique products produced by Japanese companies.

To put it simply, Japan as a country is a girl.

If countries were cars

We have seen how Japanese corporations manufacture products from a "girly" worldview. What kind of products would they make based on a masculine mentality and mature, grownup values? If such a thing as a girly way of manufacturing exists, why can't there be product designs based on a mischievously boyish point of view or marketing plans based on such male values as competition, efficiency, and rationality? How about the "fully mature woman" as a manufacturing concept?

The 1964 Mustang, a very American car.

Consider the Prius and the Civic CVCC. Who is the perfect driver for these cars, leaving their environmental benefits behind? I would argue that it's someone like the Japanese pop girls' group Morning Musume, the epitome of girly Japan. They are the perfect passengers for these softly feminine, toylike cars. You can picture them giddily enjoying the ride.

What would a manly car look like and who should be behind the wheel? Since boys tend to have a strong, childlike curiosity while openly and aggressively expressing their masculinity, a forceful, no-frills car would fit the bill.

This manly car is similar to how the Japanese see Americans: pushy, forceful, but also youthfully charming. Americans are like a powerful car driven by a beautiful

A race queen at a Japan car show.

woman. I picture a Cadillac, Mustang, or Corvette, with massive engine displacement—a Mustang convertible driven by Marilyn Monroe (though she died before the Mustang debuted, unfortunately). Or to be more up-to-date, let's put Madonna behind the wheel.

Now imagine a woman who has grown out of Japanese-style immaturity. She is glamorous and sensual, but also elegant and classy. Who would such a woman be and what car would she drive? One that comes to mind is Catherine Deneuve; another is Sophia Loren. And the car would be a Ferrari or Citroen. Both the cars and women are European.

Finally let's imagine a mature, adult man. The car for such a man would possess a stately air, be Spartan in design, and have plenty of handsomely styled functions. A German Mercedes Benz or a Swedish Volvo comes to mind. Who would be the perfect adult woman to be sitting in the passenger seat of one of these slightly intimidating cars? My German friends say Greta Garbo or Marlene Dietrich.

So if cars and women were countries, we'd have a situation a bit like that presented on the facing page. The women who best (if simplistically) represent the national character of Japan, France, Italy, the U.S., Germany, and the Scandinavian countries are perfect matches for the most famous cars produced there.

Of course, elegant cars like the Lexus and Fuga are produced in Japan while cute cars like the Volkswagen Beetle are manufactured in Germany. But if these Japanese cars were matched head-to-head with their Italian counterparts, there would be no competition in terms of elegance. Likewise, in terms of pure power, American cars are unbeatable.

Each car has its own set of pluses and minuses, but drivers who best match the image of their vehicle also look the best driving it. The same goes for the engineers of these cars. They are at their best when they can express themselves freely, instead of trying to fit their thinking into someone else's mold.

ADULT

FERRARI DINO 246

VOLVO 940 SEDAN

FRANCE, ITALY
Sporty, elegant

NORTHERN EUROPEAN, GERMAN
Sturdy, spartan

Catherine Deneuve

Marlene Dietrich

LAMBORGHINI CONCEPT S

MERCEDES BENZ G500

FEMALE ←→ **MALE**

Ayumi Hamasaki

Marilyn Monroe

HONDA CIVIC EX SEDAN

FORD MUSTANG

JAPAN
Girly, toylike

USA
Sexy, powerful

Morning Musume

TOYOTA PRIUS

CHEVROLET CORVETTE C6

Madonna

CHILDLIKE

The Prius grabbed the hearts of those who aspire to environmentally friendly glamour. However, if it were based on an American conception of glamour or had tried too hard to look elegant, it might not have been such a hit. Designed by the environmentally conscious Japanese, it naturally reflected Japanese culture, and thus became a feminine-looking product.

An interesting relationship also exists between women and military vehicles.

During World War II, Americans began the practice of painting sexy women on the noses of their military aircraft. These pictures of nude or bikini-clad girls striking seductive poses were called "nose art." The art relaxed the minds and raised the spirits of soldiers about to enter battle.

This combination of military aircraft and glamorous women is similar to the combination in Japan today of Formula 1 racing and so-called race queens—women who serve as visual decoration to the racing scene. Both reflect a male-centered world view.

Meanwhile, the world of fantasy in peaceful present-day Japan is populated by *bishojo* (pretty girls) who battle evil day and night in such anime as *Sailor Moon* and *Evangelion*. The latest combat *bishojo* are cyborgs who have fused their bodies with their weapons. Representing this new trend are the mechanized foot soldiers called *meka musume* (machine girls) designed by Fumikane Shimada.

In Japan, innocent *bishojo* fight as men traditionally do while selflessly reconstructing their own bodies and turning themselves into half-human, half-machine hybrids. This is part and parcel of the Japanese love of anthropomorphization.

Comparing the nose art of America and the combat *bishojo* from Japan, we see more clearly what the two countries have in common: Both are childish, generating art forms unlikely to emerge from Europe's mature, traditional culture.

chapter 4

"Taste" from a Computer-Like Brain

Processing the sounds of nature as language

Why are the Japanese so girlish? Explanations for this cultural disposition can be found in everything from biology to history. One is geopolitical: The Japanese have always lived on an archipelago where the threat of an attack from a neighboring nation is almost nonexistent. The Japanese also lack a sense of crisis due to Japan's renunciation of war after World War II.

Another is a theory of evolutionary biology called neoteny. According to this genetic-based theory, Mongoloid peoples, of which the Japanese are one, have relatively underdeveloped masculine traits, leaving them with an infantile character. The cause is said to be related to hormonal secretions.

Some also say that the character of the Japanese language encourages "feminine" ambiguity. As a result,

Japanese speech, especially that of women, tends to be indirect and wordy, if quite polite. For example, a female Japanese cashier might say, "If you would like a receipt, I'm afraid you will have to wait a moment. Is that all right?" This is called "polite part-time worker" diction.

Whenever I hear such a phrase from the mouth of a young part-timer, I find it hard to believe those who say young people nowadays want to be more Westernized. Instead, the ambiguity of their speech indicates that they want to be more purely Japanese.

One particularly interesting theory has been proposed by Dr. Tadanobu Tsunoda, a neurologist and honorary professor at the Tokyo Medical and Dental University. In research he conducted in Montreal in 1978, Dr. Tsunoda discovered a brain mechanism found only in people who speak Japanese and Polynesian languages.

Using the Tsunoda Key Tapping Machine, an analog device that measures the switching mechanism in the brain stem, he came up with a method for determining which half of the brain is processing a sound at a given moment.

Through repeated testing, Tsunoda discovered that people who had learned either Japanese or a Polynesian language as their mother tongue between the ages of six and nine showed a greater development of the left brain, which controls the language process, regardless of race, nationality, or ethnicity.

Both Japanese and Polynesian speakers rely primarily on vowels to determine the meaning of a spoken word, while people who speak other languages depend more on consonants. Also, vowel sounds are closer to the sounds of nature than they are to human voices.

Japanese people thus process the sounds of nature—the twittering of birds, the susurration of falling rain, the moaning of the wind—with the same side of their brain (the left side) that processes language. People who do not speak Japanese or Polynesian languages process speech with their left brains, the sounds of nature with their right. That's why the Japanese believe that meaning is being conveyed even in the noises of insects or the babble of a brook. Similarly, communication in Japanese is often more of an exchange of emotion than information. To those whose mother tongue is Japanese, conversation is akin to the growling of an animal or the tweeting of birds. This communication style emphasizes feeling rather than content, just as a song does.

If Japanese indeed lacks logic compared to other languages, the reason may lie in the different way the brains of Japanese and the speakers of other languages process sound.

Left-wing overseas; right-wing in Japan

Tsunoda's research implies that the brain is like a computer, and learning a language is similar to initializing the electronic circuits of a computer with an operating system. At the heart of Tsunoda's research is the observation that, when the brain of a child is initialized with Japanese or a Polynesian language, its modes of response differ from those initialized with other languages.

For Tsunoda, anyone who learns to speak Japanese as his or her native tongue between the ages of six and nine is Japanese, regardless of race or background or DNA. The international reaction to Dr. Tsunoda's research indirectly proves how correct it was. Many in Japan who adhere to a nationalistic ideology—the so-called right-wingers—welcome Dr. Tsunoda's theory. They believe it proves that Japanese are gifted with special abilities by virtue of being born Japanese. Dr. Tsunoda has never said his research supports any theory of Japanese superiority, but it has often been interpreted that way.

On the other hand, left-leaning Japanese reject

"Taste" from a Computer-Like Brain

Tsunoda's theory in principle. They can't accept that Japanese may be unique or different from other ethnic groups. When they hear a view implying that Japanese are special in some way, they shut down their brains and stop thinking. Furthermore, they have labeled Tsunoda a nationalist and tried to get him expelled from academia.

Interestingly, the reaction overseas has been the opposite, with academics in (the now former) communist states and the developing world showing the most interest in Tsunoda's theory. They have interpreted his research sympathetically, based on logic instead of emotion.

Tsunoda's theory of brain initialization through language has nothing to do with such factors as genes or race. It does not undermine the socialist ideal of human equality. It also does not discriminate against poor people in developing countries. Tsunoda's theory has been placed in a strange position: Foreign socialists have supported it based on logical reasoning, while its Japan adherents are mainly nationalists who defend it emotionally. This could be taken as ironic proof that Japanese process language (or rather language theories) through their emotional left brains; non-Japanese, through their logical right.

chapter 5

Otaku Manufacturing Is the Key to Japan's Future

Manufacturing education at its best

The Japanese national character—the most "girly" in the advanced industrialized world—was quite different not long ago.

From the Meiji Period (1868-1912) until the Pacific War (1941-45)—a period when Japan had to compete with the Western powers in the economic and military spheres—manliness was strongly emphasized. Vestiges of this attitude could be seen in the gung-ho days of rapid economic growth after the war and the go-go Bubble economy era of the late 1980s.

During the "manly" prewar era, the catchwords for Japan were "cultivation of industry and promotion of business" and "rich nation, strong military." During the war, the slogan was "one hundred million hearts beat

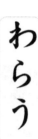

Hiragana, *created by women in the Heian Period, has ties to today's emoticons. The word at right is* warau, *"to laugh."*

as one." In the Bubble era, the question posed by a hit energy drink commercial was "Can you fight twenty-four hours a day?"

Compared with those eras, today feels like a revival of the generally relaxed and peaceful Heian (794–1185) and Genroku (1688–1704) periods. The Japanese are entering a period of fully ripe "girlishness."

The *kana* (Japanese syllabary) culture invented by Japanese women in the Heian Period nearly a thousand years ago greatly enriched later Japanese culture. In a kind of imitation, the *gal* of the Heisei Period (1989–) have invented a culture for expressing their true feelings in between the lines of their everyday lives. Their use of emoticons is one example of this.

Japan has succeeded in becoming a prosperous

country with its long-standing policy of catching up with and passing the West. But as long as Japan continues to work within the borrowed framework of a Western culture, with its concepts of "adult" and "manly," while prioritizing the production efficiency (think of military weapons as a prime example), it will have trouble being anything but a follower. It may lose the world's respect as well.

Instead, Japan should confidently promote manufacturing that consciously and fully uses its geeky and girly temperament, as clearly seen in the *otaku* and *gal* cultures. Such a movement has already begun, as Japanese makers achieve great results with machines that respect people's individuality, are considerate of others, try to make the world a better place, and bring people closer together.

Just by virtue of being raised in Japan, speaking Japanese, and living comfortably in this country's ripe girly culture, the Japanese have received a world-class education in rational, high-quality manufacturing. I believe the future will be bright for Japanese manufacturing and the country that supports it.

The story of the 2045 Japan-American War

Japanese anime and manga are already spreading around the world. Taking note of this trend, the Ministry of Internal Affairs and Communication has drafted a digital contents strategy.

The popularity of *Pokemon*, *Sailor Moon*, and other Japanese anime abroad inspires both hope and joy in me. Ichiya Nakamura, professor of the Graduate School of Media Design at Keio University, told me a wonderful fictional story I would like to share with you.

The place is Washington, D.C.; the year, 2045. In a conference room in the Pentagon, government and military leaders are engaged in a heated debate.

The reason: After nearly a century, Japan and the U.S. are again at war. The debate in the conference room is over whether the U.S. should attack the enemy capital, Tokyo, with weapons of mass destruction. The war situation is becoming critical, and the participants are being driven to make a decision. A boisterous contingent is saying, "Attack Tokyo!"

In the midst of this commotion, a high-ranking official mutters, "Are we going to incinerate Sailor Moon's sacred ground of Shirokanedai?"

Hearing this, the other officials look at each

other, taken aback. They too were once children crazy about *Sailor Moon*. The outcome of the meeting is not important. Even if the Americans attack Tokyo, they will feel a lingering sadness and guilt. That's what's important.

If Japan's anime culture spreads even further, it can link hearts worldwide and make a major contribution to peace and mutual understanding. That is also Nakamura's dream, I believe, put into the form of a story.

It's wonderful that the world's children are enthusiastic fans of anime. Their understanding of Japan's girly culture will expand over the course of a generation. A new breed of young, foreign Japanophile is emerging. The new fans of Japan won't be Orientalists, but they will be anime-savvy. When I talk with these Japanophiles, they nearly all say that they became interested in Japan from watching anime when they were little.

Tapping latent talent

In his 2007 book, *The Loneliness of Freedom and Prosperity*, former Prime Minister Taro Aso writes about the far-reaching effects of anime and manga culture.

The Japanese soccer anime *Captain Tsubasa* has been translated into various languages, Aso notes. In these translations, the characters have been given localized names while becoming idols to young soccer fans everywhere. Aso writes:

> The world does yet understand the value of comics . . . When asked "Why did you start playing soccer?" Francesco Totti of Italy and Zinedine Yazid Zidane of France both say, "*Captain Tsubasa.*" Zidane, who made France No. 1 in the world, read a Japanese comic called *Captain Tsubasa* and decided to start playing soccer.

It's well known that Zidane, former captain of the French national team, is the son of Algerian immigrants. In the Middle East, Tsubasa, known there by the Arab name Majid, has even had an impact on the region's tense political situation. Allow me to quote from Aso's book again:

> Water supply trucks sent with the Japanese Self-Defense Forces troops to Iraq did not have the Japanese Hinomaru flag on the side, but rather a picture of Captain Tsubasa. That's why they were never attacked.

In other words, Captain Tsubasa protected the Self-Defense Forces from harm in Iraq.

Aso understands the true nature and power of subcultures. He is the rare Japanese political leader able to use that understanding in dealing with the world's power brokers.

I'm glad that the government's digital contents promotion policy is raising anime's status and strengthening its role as a pivotal industry, but job creation remains a difficult problem. Even if you have talent, it doesn't mean you can succeed as an anime creator.

As many as thirty thousand manga "circles" (groups of manga creators) sell their wares every year at Comiket at the Tokyo Big Sight event space. These self-publishers have a surprisingly high level of skill. In Japan, it's impossible to succeed as a manga artist or animator with average talent.

Japan's leaders ought to make better use of these many creators in labor-intensive industries. It's important to develop products using anime methods and mindsets.

If Japanese manufacturers make this a reality, following the ten "rules" outlined in Part 2, they can not only maximize the talents of these creators, but open up new opportunities for themselves. Then Japan can, for the first time in its history, aim for the economic summit on its own, without following in the footsteps of others.

Conclusion

Japan's original tools and products are praised overseas, but at home, they are considered the norm. There have been many analyses of the Japanese and their culture from the standpoint of linguistics, psychology, anthropology, and sociology. I may be the first, however, to make such an inquiry from the viewpoint of manufacturing.

Astonishing new Japanese products continue to appear from creators with deeply original minds and ideas. If you discover such products please contact me by email at m@morinoske.com. Your suggestions will be used in my research and writing.

Japan's intellectual class tends to look down on assertions of Japanese uniqueness. When anyone talks about the special talents of the Japanese, some people become strangely irritated. If I may be a little sarcastic, this is an example of "Japanese modesty" that has few parallels elsewhere.

But like it or not, Japan accomplished a miraculous recovery after the war and became one of the world's leading manufacturing countries. This alone indicates that the Japanese possess special talents. And Japan's geeky and girly culture, as exemplified by the *otaku* and *gal* subcultures, is definitely out of the ordinary.

Japan has the potential to be a world leader in the fields of industrial technology, politics, and culture. The Japanese can take pride in their geeky and girly culture, which has given birth to so many only-in-Japan products.

Upholders of traditional Western values have long viewed Japan as childish and irrationally immature. But now times are changing. Now the world is focusing on geeky and girly culture, in which Japan is a world leader. Manufacturing innovations follow in that culture's wake.

Of course, I understand the feelings of those who complain about the indifference to the past shown by many young Japanese. Some don't even know that Japan and the United States fought a war. I sometimes want to shout, "Who made Japan this way?"

I also agree with those who argue that the Japanese educational system should be overhauled. The government needs to reform it from top to bottom.

But the Japanese should not reject *otaku* and *gal*

culture. Even though its shallowness may anger us, we should calmly view its potential with the mind of a parent thinking of its child.

Children need to be corrected, but they also need to be cherished and have their talents nurtured. Japan needs to nurture, not crush, the *otaku* tendency to obsess over details as well as what looks be their girly weakness. We should use their talents to improve society as a whole.

We should not reject *otaku* just because we don't understand what they are thinking. They excel at coming up with original answers to the question of "What shall we make?" We need that ability to promote Japanese industrial growth.

We also have to support the digital-contents industry. However, it alone can't generate enough wealth for the happiness of the population as a whole. Instead, we should apply the superlative talent found there to the manufacturing businesses that are the pillars of our economy and fundamental to Japan's revival.

In this book I have used the expression "girly" (*onnanokoppoi*) to describe the Japanese character.

Japanese have often used another expression, *onna-kodomo* (women and children), as in "This is no place for women and children." It's an old-fashioned word with a nuance of discrimination against women.

Perhaps the true, Japanese-style "place for women and children" is finally coming. This time, it won't be discriminatory or sexist; this time, it will embrace girly and childish traits—they will become the engines that drive Japan's economy in the future.

An old saying puts it best: "Know your enemy and know yourself, and you will survive one hundred battles." When we realize that our girly and childish qualities are our strengths and strategically create products based on them, we can make great economic strides and earn the respect of the world.

But a girly sensibility alone will not necessarily create great products. People in the arts world, a tough environment in which creativity is demanded, often tell me that young people are overflowing with ideas, but many of those ideas are shallow and lack universality. The art based on them is insubstantial. On the other hand, when these young artists become older, their creativity dries up, but since they are now veterans in their field, they know how to give their ideas form.

Manufacturers have to understand thoroughly the sensibility of the young as symbolically expressed by the

Akiba *moe* and Shibuya *gal* subcultures and give that sensibility form in outstanding products, made efficiently and well.

In other words, they need to serve as a bridge between the young and the production front lines.

This lack of bridge building has profound impact on the nation. Too many young NEETs (not in employment, education, or training), *hikkikomori* (youths who have become reclusive and withdrawn), and even young runaways, both boys and girls, have talents that are never given a chance to shine. That we effectively discard these youths is our great loss. I hope that politicians and businesspeople will realize the importance of this problem, build bridges to these youths, and draw forth their hidden power.

The conventional wisdom in Japan these days is that the unique aspects of Japanese products are liabilities in the global marketplace. The thinking goes: Japanese manufacturers focus on special, luxurious functions demanded in the domestic marketplace, but they ignore international best practices. Their products have limited global appeal and have earned the nickname Galapagos because they are like the rare flora and fauna found on that island.

Conclusion

This is mistaken logic born from Western-style reasoning. Japan should not trouble itself with this specious argument.

If the Japanese keep sight of their true nature as a culture and continue to make products based on that culture's strengths, they will be richly rewarded with new sources of wealth.

Acknowledgments

I would like to warmly thank Taro Aso, the 92nd prime minister of Japan, who took time from his busy schedule as foreign minister in 2007 to read my manuscript and give me valuable suggestions.

I would also like to thank Yuko Kanazawa of Arthur D. Little, who labored tirelessly in the preparation of this book.

I am also greatly indebted to Kazumasa Ashihara and Azusa Umetsu of the same company, as well as to Jun Hirobe of Kodansha and freelance editor Natsuko Kobayashi, for the editing of this manuscript.

Huge thanks to Peter Goodman, publisher of Stone Bridge Press, for believing in my book and so lovingly creating the English version.

Also, thanks to Mark Schilling for his expert translation and Bruce Rutledge for his editing expertise.

More than anyone, I would like to thank my wife

Judit Kawaguchi for continuing to supply fresh insights and energy for my writing.

Morinosuke Kawaguchi

IMAGE CREDITS

About the author

MORINOSUKE KAWAGUCHI is a Japanese technology consultant working for the global management consultancy Arthur D. Little, Japan, Inc. He advises Japan's largest corporations in the automotive, electronics, and telecommunications industries on their strategies in both R&D (research and development) and MOT (Management of Technology).

He is also a bestselling author whose approach to subculture and how it provides an R&D advantage has made him popular in Japan and around the world. An expert on how to create innovative products with a competitive edge, Kawaguchi is an acclaimed international lecturer on subculture engineering, emotional design, and industrial strategy.

Geeky-Girly Innovation is the English translation of *Otaku de onnanoko na kuni no monozukuri*, which was published in Japan in 2007 to rave reviews. In the following year the book received the prestigious Nikkei BP BizTech Book Award for its contribution to the advancement and development of technology and management. It has been translated into Chinese, Korean, and Thai. Taiwan's Small and Medium Enterprise Administration in the Ministry of Economic Affairs and South Korea's Korea Institute of Industrial Technology have studied its concepts and recommended them to be applied in the development of their countries' technology and innovation strategies.

Kawaguchi's second book, *Sekai ga zessan suru: "Made by Japan"* (The World Acclaimed: "Made by Japan"), was published in December 2010.

Morinosuke Kawaguchi's website is **morinoske.com**, and his videos can be seen on the YouTube channel **JapanTechLessons**.